The W

The universe: what it is and what to do about it

Max Merrybear

Published using Amazon's CreateSpace by Dylan L Garda Virtual Publishing, Salisbury, United Kingdom

ISBN-13   978-1478139331

Copyright © Max Merrybear 2012, First edition.

Max Merrybear asserts the moral right to be identified as the author of this work.

This is the UK English version, so some words might look a bit funny if you're not used to the written language of Her Britannic Majesty, such as 'twee' and 'cheque'. We get US English books quite a lot in the UK. Some people moan about it. There are probably worse things to moan about though. Whether or not this book ever gets translated into US English or even Australian English remains to be seen of course.

All rights reserved. That said, you're unlikely to burn in the fiery pits of hell if you lend it to your grandmother to read, nor shall you be plagued by sulphurous demons if you donate it to the book stall at your local church bazaar and they sell it for 20p. Please visit www.maxmerrybear.com for permission rights if you wish to quote over 200 words; otherwise, as long as you've read this book in full and your conscience is clear, just feel free to do it.

At least 10% of any profits from this book will be donated to the Dogs Trust (www.dogstrust.org.uk); not that many dogs make a profit out of publishing once production costs and distributor commission are deducted. Other beneficiaries are mentioned later in the book. I have no great need for money myself because I'm a dog.

Dedications and Thanks

With thanks to you dear reader, as long as you bought or borrowed this legitimately or were given it as a present, rather than pinching it, I hereby dedicate this copy to:

....
(Please insert your name above ... which may be awkward if you're reading this on something electronic ... or you don't own a pen ... or you haven't got opposable thumbs etc ... Sorry.)

Also by Max Merrybear

Actually, nothing at the moment. Early next year I shall help my platonic human male life partner, Mad-B, publish his work of fiction *A Pat on the Head* to be followed by *The Beaujolais Society*. See www.dylanlgarda.com for more details of those.

I'll keep you posted about my future ventures, literary and otherwise, on www.maxmerrybear.com, but it might take a while depending on how Mad-B's above works grind through the publishing mill. Life can be quite busy when you're a dog!

About the author

Max Merrybear is a dog. If you've read Chapter One you'll already know that. If you haven't read Chapter One yet, now might be a good time to start.

Contents

The Word of Dog is written to be savoured patiently and conscientiously from beginning to end, so for the first time you read it a contents page would seem somehow inappropriate. For ease of future reference, if you wish, feel free to write a personalised one here. Unless you're reading the Kindle version, in which case you probably know how to tag stuff ... Not that I'm too sure because of the lack of opposable thumbs, which makes a Kindle quite awkward to use ... Sorry ...

Chapter One

Introduction

So, who are you? A scientist? A theologian? An ordinary everyday person trying to make sense of it all? A bit of each? Me, I'm a dog: a 12½ stone Newfoundland, Max Merrybear, black and hairy, son of Champion Watervalley Marvin Gaye of Merrybear. And Gemma.

And recently, I died. Not for long, admittedly, and not as dramatically as it might have been. I'm certainly not claiming a divine post-crucifixion day-three resurrection. But died I did. If this were a folk song I'd be singing, 'Died I did diddly I.' But this isn't a folk song, it's a book. You probably already know that. However while I was dead, albeit briefly, I did learn a few things that might interest you. Things that you might not already know, or perhaps you might have a nagging doubt about whether the things are as clear-cut as they should be. Things such as:

- The universe: what it is and what we should do about it.
- Exactly what might possibly happen after death, especially to the pets and people you've loved.
- Why life isn't always heaven on earth for everyone, not even humans.
- Whether science and seemingly conflicting religions can ever be concurrently correct and, if so, how.

- What you might decide to believe if you'd rather things weren't quite the way you currently see them.
- What you might decide to believe if you think a dog might make more sense than many of the people you meet.
- What to do if you find a famous artist standing in a bucket on your kitchen table.
- Whether there's a Father Christmas and, if there is, exactly how he overcomes his more obvious logistical problems to go about his business.
- Why whatever *you* decide to believe by the end of this book, not what me or anyone else tries to tell you, but what *you* personally decide to believe is *genuinely* the way things *actually* are.

This, then, is my tale. That's a pun. Get it, tail? Remember, I *am* just a 12½ stone puppy, albeit a very old, hairy and wise one.

So. Let's get straight down to business, because I know what you may or may not be thinking, you may or may not be thinking, 'What the heck?!' or at least words to that effect, and I can't say I blame you. After all, it's not every day a dead dog comes up with the answer to life, the universe and everything, then spends a year pawing over grammar and punctuation textbooks to hone his previously rudimentary literary skills, and presents his findings in a well-reasoned and eloquent tome; especially as the book includes answers to the forty-two most salient questions anyone is ever likely to ask but then, as I'll explain later, today isn't quite like any other day (even though for many of us it probably very much is, but we shall discuss that more in a later chapter as well).

Naturally I also asked a few of my human chums to read the manuscript and make comments before it went to press, including a number of eminent scientists and theologians. Four of them have science doctorates and are pushing back the boundaries of stem and fuel cell research, while another has acted as an adviser to several bishops and even the Archbishop of Canterbury on ecumenical theological matters so, in terms of credentials, you'd probably have a dog-gone hard time finding a more commendable puppy. None of them disputed the manuscript's contents, but I felt better for asking, especially as I got a few tummy rubs out of it. It would, however, be remiss of me not to mention that rather than to appraise my writing, the *actual* Bishop mentioned on the back cover who said, 'Who's a clever boy then,' spoke the words in what has to be said was a disapprovingly 'cheeky' manner when I, ahem, covertly sampled his mushroom and goat's cheese risotto during a somewhat protracted pre-supper grace … I almost wish I were joking, but I'm not … Sorry …

But before we get going properly, I thought you might like to know a little bit more about my life, just for the social context, and because then the occasional references to some of the humans, dogs, cats, and even the horse in my extended family will make more sense. I will be brief though; the last thing I want this to be is my autobiography. That will probably come later once I've spent the royalties from *The Word of Dog*, mainly on a number of worthwhile projects we'll cover later, some of which may well be of interest to you, particularly if you like to travel.

First of all there's My Adoptive Dad Bob, or Mad-B for

short. He's been my human life partner since I was nine weeks old, although I'd met him twice before that, initially when I was just three weeks and six days old. To say that we have a special bond would be a huge understatement, despite the obvious differences in our intellect: Mad-B never was the brightest bone in the shrubbery, but I love him regardless and I know he loves me. You'll hear more snippets about him as and when required, although for now suffice to say that I think he feels his age a bit and, while he's large like me and hairy in a lot of places, his fur is going rather thin on top.

I love his long-suffering wife Mrs Mad-B as well, although I've only known her for a little under three years since they first met. She thought, as I was immaculately groomed when she saw me out walking, that Mad-B would have an adequately clean house and mind as well. In reality he had actually just paid around £240 for a protracted course of professional knot removal, dematting and grooming for me, which meant he was feeling skint, so on their first proper date they had to split the bill. And he pretended he couldn't possibly manage dessert. And she yelped in horror the first time she saw the state of his cluttered, dusty and hairy house. I think she's learned a thing or two about making assumptions now, although she tells me she's still really glad she met him.

Jazz is my 'cognitively challenged' stepbrother, although he's fun to have around. He's a Labrador cross, although we don't know what he was crossed with as he was found starved and tortured on the streets of one of the less celebrated high-rise estates in Glasgow, eating bones from the gutter to try to survive, before being shipped south to a Dogs Trust in Wiltshire for rehoming. We think he's just over three years old now, although we can't be sure, and he does still have a few issues, albeit nothing too serious these days. Except the one I dare not mention. But don't worry, I'll mention it later. Several times.

Apart from that, Mad-B and his wife also brought a couple of adult kids each into the equation, one of whom, Abi, still lives with me, and a grandson, three cats and a horse. Mad-B swears blind that the horse has laser vision and breathes fire from her nose, although I have to say she seemed perfectly charming when I met her.

Oh yes, and when we're out walking, Mrs Mad-B goes a little bit loopy every time she hears, 'Gosh, I thought it was a bear!' And as any human with a large Newfoundland life partner will tell you, this can happen several hundred times a day if you go anywhere crowded. So if you feel like playing a little game, why not say, 'Gosh, I thought it was a bear,' every time you see the word 'bear' in this book. You don't have to of course: as you'll soon discover this book is all about you having the freedom to make your own choices and to be confident that you're right, but by all means *bear* it in mind.

OK. That's enough idle chat. I expect you want to know a bit more about the nature of the universe. Luckily, we cover this in more detail in Chapter Two, cunningly entitled *Chapter Two: The Nature of the Universe*.

But before we dive in, there's one thing you ought to know: the universe is *weird*. Weird almost beyond imagination. So what I intend to do is explain things really simply, but also have some more complex yet free-standing paragraphs to keep my credibility among the nerds: Trekkies, Mathematicians and assorted types of Physicists for example. I'll highlight the bits, many of which are in Chapter Two, that you, a normal person, are welcome to skim over or ignore completely by introducing them with '**Nerd Alert!**' on a separate line and in bold. Afterwards, to show the end of the nerd alert, we'll have '**Back to normal**'. And just for a bit of extra fun and excitement, we might even have some similar slightly wackier introductions here and there to set the tone.

So remember, you're free to ignore nerd alerts. Unless you're a nerd. Or maybe just nerd-curious. Simple? I thought so. Let's give it a try. This one's bigger than most. If you're normal, I can only apologise, especially as we barely know each other yet, and please remember that you're completely free to skip to the paragraph below where it says '**Back to normal**' on a separate line in reassuringly bold letters. On the other hand, if you're a nerd, enjoy!

Nerd alert!

Dear Nerd. In this book I shall mostly refer to the universe as having three spatial dimensions: height, breadth and width. X, Y and Z if you prefer. I am well aware that nerds regard spacetime as *highly* curved/warped such that a quasar which emitted light 12 billion years ago (with all due deference to those who believe that the universe was created 6,000 years ago), and where that light travelled to us at the speed of light, is actually 28 billion light years away rather than 12 billion, and with the current diameter of the observable universe now usually quoted at circa 93 billion light years with a minimum of 78 billion, rather than just 27.4 billion which it would be if it had expanded at the speed of light since the alleged big bang. For normal brains this is stupid and unnecessary. And as much as I love science, and I am generally somewhat nerdy myself, if in your theorems you use words like 'infinity' or phrases like 'ignoring radiation in the early universe' deep down you must know you're probably getting *something* wrong. So in this book we mostly have just three-dimensional space for ease of discussion. Convert it to four if you want to, but this is a book mainly for normal people. And remember you might not be right, although as we shall see I'm happy to believe that you are, so please try to keep an open mind, as indeed shall I. As justification for asking you to keep an open mind, I cite the case of Newton's Corpuscular Theory, which was at the very forefront of intellectual genius until later generations realised that Newton had not only got it wrong, but that his gravitas had held up Huygens' work which was a whole lot more accurate! So, to summarise: this book, three dimensions; your head, four dimensions. Let's get on with the next chapter. I think you'll like it. It's got eleven dimensions ...

Back to normal.

Right then, onwards!

Chapter Two

The Nature of the Universe

A tricky subject, I think some of you will agree, while others will disagree on the basis that it's perfectly obvious what the universe is, whether or not there is just the one universe, and for how long it has been around. In terms of how long it has been around, we'll leave the discussion about what *time* is and what, if anything, to do about *that* for a later chapter. For now it can just be something we get from a clock.

What we're going to do in this chapter is some of the donkey work that will, by the end of the book, allow us to understand the universe fully with no remaining gaps or doubts. However, as not everyone thinks that donkeys are fun, and I do very much mean 'by the end of the book' rather than 'here's a quick-fix answer in short sentences with words of one syllable', I'll also throw in a handful of Mad-B's hilarious but random embarrassing personal secrets. I hope that'll be enough of an incentive to keep your tummies rubbed. Feel free to rub mine as well by the way. As you may have inferred from Chapter One, I like that.

Luckily as we, that is to say both of us, me and you together, mostly you yourself in fact with just a little bit of help from me here and there, shall attempt to prove beyond any doubt whatsoever, the universe both is and is not tricky. However I am confident that you yourself, ie the person, dog, dolphin or other sentient being actually reading this text, will almost certainly find the universe *far* less confusing by the end of the book, however weird or otherwise it may seem at the moment. I can't promise that it'll be any less confusing by the end of this *chapter*, and as implied I make no such claim, but as long as you're happy to read *The Word of Dog* patiently, sequentially and conscientiously from beginning to end, feeling at least a comfortable level of understanding at each stage, even if you ignore a bit of the maths, the wonders of the universe should unfold by the end of the *book*.

Clear? I do hope so; that was a bit of a mouthful, although just in case it helps, maybe I should paws (sic – it's a dog pun – sorry) for a minute to let you have another read of the previous three paragraphs … OK … Well I think that (simplistically) illustrates a point: time is weird. I didn't actually pause at all. I actually wrote this chapter quite some 'time' ago. I'm doing something completely different now. What, I have no idea, mainly because there's no real way for me to guess when you might be reading this. Not that either of us need to worry about that, especially not at the moment when we've still got a whole universe to discuss. Starting now. Whenever that is.

Naturally, being a wise old Newfoundland, I have sufficient clarity of thought and insight not to make generic assumptions about how best to help your own personal understanding. Therefore, as a gentle introduction, I shall propose five pretend models for the universe, and you can just choose the one that makes most sense to you today to help you visualise our discussions. Tomorrow you can choose a different model if that helps. It's your life. We'll start with something *really* simple and work our way up to something rather more, ahem, complicated. And if maths makes you squeamish or bad tempered, by all means skim over any bits that offend, although most of it wouldn't even get you a GCSE or High School Diploma; the worst of it will be in 'nerd alerts' anyway, so it's largely just there to keep the nerds happy rather than being essential for any normal right-thinking individual such as yourself.

I shall start with a model that even Jazz could understand, or at least he'd be able to understand it if he could stop rooting for truffles on the kitchen floor for a few seconds. He is a *bit* tubby, but then he is mostly Labrador, so naturally he always thinks he's starving, which is certainly no longer the case even if it once was.

a)   The Simple Appleverse

If you're a visual sort of creature you may like to get some props. Otherwise you can just imagine it; this one really IS simple. In fact I rather hope you don't think it insults your intelligence. Maybe you could think of it as a warm-up exercise for your brain, just to get into the swing and to start us off on the same path? For this explanation you'll need at most four apples (or things you can pretend are apples), one coin that either has a 'head' on one side and a 'tail' on the other, or things that you can pretend are heads and tails, and a piece of paper with four squares drawn on it, each large enough to take one apple, and numbered one to four, as per the diagram below (only a bit bigger unless you have really small apples):

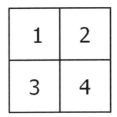

Simple? Good. I'll just paws for a minute ... No, just kidding ... It's that *time* thing again ... I probably won't pause at all, although you are entirely welcome to of course ... and I promise that I hope the jokes will probably improve soon with a bit of luck once we've got to know each other a bit better ...

Now, if you would be so kind, please toss the coin four times. If you *won't* be so kind, not even by way of pretending to do it, or at least acknowledging that you could do it if you felt like it, we may have a problem, so I hope you'll BEAR with me and indulge me for now.

For each 'head' put an apple in the corresponding square, and for each tail (wagging or otherwise) leave it empty. So if you got head, tail, tail, head, put an apple in the first square, leave squares two and three blank, and put an apple in square four. Get the picture? You do? Hurrah! Now **bear** with me a bit longer. How about if I suggested that what you've done, given that this is what we said we'd do, is make a model of the universe, and that in your model you have two lumps of matter (or 'stuff' if 'matter' sounds offensively scientific for your current disposition) and two lumps of empty space? Simple enough as a concept? Call it art if that helps? No desire to murder anyone or have a nervous breakdown yet, or at least no such *new* desires: in a much later chapter we'll cover what you might wish to do if you've had any recent murderous desires or a nervous breakdown at, for example, work. Splendid!

Now get your best ever friend, spouse or partner to have a go, or if they aren't available just picture them in your mind and take their turn for them. If you've never had a best friend, spouse or partner, you could fetch your favourite teddy bear and pretend to take his go for him. I say 'his' because I've never known a favourite bear who wasn't called Edward, although I confess my experience of bears is actually extremely limited, so if you've never had any friends I suppose you might have an Edwina Bear: anything's possible, and anything *is* possible, as we shall see later! If you've never had any friends, acquaintances or childhood mannequins of any description or designated gender, I'd really suggest that if at all possible you go out and get at least one of the above before you go completely insane, although for the sake of expediency here, if you really have no alternative, by all means use a carrot, draw a little face on it and call it Mrs Bananahead.

Anyway, back to the apples, and for brevity let's just call the person or other entity your 'friend' and assume that they get head, tail, head, tail, and so their pretend model of the universe looks pretty similar to yours, albeit with the apples in slightly different places. It's probably not worth poking them in the face with a sharp stick because of the subtle difference, is it? If you *have* answered 'yes, I want to poke my best friend in the face with a sharp stick because I find that the position of their apples is offensive to me,' it might be worth consulting an anger management and/or relationship specialist before reading on (hopefully not by necessity a relationship specialist who deals with carrots named Mrs Bananahead, but if you just stabbed a carrot because you were offended by some apples I certainly think it might be worth you speaking to *someone*), otherwise great: we're making progress!

In fact, looking at just *your* Simple Appleverse and ignoring your friend's version for a moment, because you tossed the coin four times, and each coin could have landed one of two ways, there were 2 x 2 x 2 x 2 possible Simple Appleverses, or 16. If you have mildly nerdy mathematical tendencies you may remember from primary/elementary school that you can call this 2 to the power 4 if you like, or 2^4, or 2^4, or 2**4 in Rexx on an IBM Mainframe, however you like to represent it really, but at the end of the day, 16 is what it is.

Nerd alert!

Or in Excel you could call it '=EXP(4*LN(2))'. And by the way, if any *really* pedantic mathematicians are reading this, for the purposes of our discussion if the coin ever lands on its side or rolls unseen under the sofa it is always re-tossed to give a satisfactory head/tail result on a precise 50:50 unrounded true/false Euclidean basis in three dimensions only with no unresolved Heisenberg uncertainties. However, the accuracy of the coin toss is a valid point worth any nerds bearing in mind; I certainly don't want anyone left feeling dissatisfied by the Simple Appleverse, nor indeed confusing it with the Grand Appleverse we'll discuss a bit later!

Back to normal.

For any non-nerds, and I've inferred quite possibly that's most of you, who happened to scan the above nerd alert out of curiosity and saw the words Euclidean and Heisenberg, here's a little nerd joke by way of explanation. If you want to skip this, just fast forward to the 'End of nerd joke' in bold below. Before the joke, as background information, I should probably mention that Dr Werner Heisenberg was a chap who all the way back in 1927 proposed that for really, really small things like electrons (bits of electricity) or smaller things still, the more accurately you measure the position, the less accurately you can measure the velocity, basically because the act of measuring does 'things' to the system which affects one or other of the values. This gives rise to the following hilarious joke:

Nerd joke!

A policeman stopped Werner Heisenberg for speeding in his car and said, 'Do you know how fast you were going, sir?'

Dr Heisenberg replied, 'No officer, but at least I know where I am!'

That's quite funny if you're a nerd. I do realise that the jokes still need to improve though, and soon ...

Euclid was an Ancient Greek who is largely responsible for Euclidean geometry, ie the geometry that you probably got forced to do at school. Non-Euclidean geometry is the mathematical study of, among other things, what happens when parallel lines crash into each other. It's completely useless, apart from the fact that if you're a nerd that's allegedly what might actually happen in four-dimensional spacetime. There are no geometry jokes. Geometry isn't fun and it certainly isn't funny ... unless perhaps you're a really special sort of nerd.

End of nerd joke!

Now then, for argument's sake, how about if I suggested that we pretend that you and your best ever friend have just made models of the *same* Appleverse, one second apart, and that what we've done is to show how one single apple 'lived' in the Appleverse, a bit like a cartoon on television where each successive drawing/cell/frame is slightly different to the next one and therefore shows how things change from one (split) second to the next? And how about if we say that any difference in the position of the apple is just because the outcome of something happening, ie the coin toss, differed from one second to the next? I hope that sounds reasonable, given that you just did it. Now what we have are sixteen possible Appleverses one second, and sixteen different possibilities the following second, or 16 x 16 possibilities, or 256 possible combinations of Appleverse altogether, some of which no doubt make a lot more sense than others!

Pause for reflection if you need to. Breathe deeply and muse if you like. It's worth picturing and feeling happy about it. Likewise, if 256 sounds like far too large a number of possible different combinations to believe, you can always get a sheet of paper and write them all down thus:

1) Tail, tail, tail, tail   and tail, tail, tail, tail.
2) Tail, tail, tail, tail   and tail, tail, tail, head.
3) Tail, tail, tail, tail   and tail, tail, head, tail.
4) Tail, tail, tail, tail   and tail, tail, head, head.
... and so on until ...
256) Head, head, head, head   and head, head, head, head.

Or you could just take my word for it. It's up to you, although whether through blind faith, simple maths or tortured experiment, it *is* quite important for the purposes of our discussion that one way or another you believe there *are* 256 different scenarios!

Now let's look at a couple of examples. Say you and Mrs Bananahead have both thrown all heads: in one second your Appleverse is completely full of apples, and the next second everything is the same as it was and the Appleverse is still full of apples. It'd be a fairly dense and crowded place to be, but at the same time perhaps calmingly stable in its consistency. Similarly, if you've both thrown all tails, your Appleverse would be all cold, lonely and bereft of apples for both seconds of its existence, although still stable in its emptiness.

Alternatively, you may have thrown head, tail, tail, tail and your friend tail, head, tail, tail. What we now see is a rather splendid Appleverse, still not to be confused with the Grand Appleverse, however grand it may now look, with one apple that starts in square one and moves serenely over to square two during the course of a second. Yippee, is that fun or what?!

Then there's the dark side: head, tail, tail, tail followed by tail, tail, tail, head. What's happened now is that the apple has started off in square one and a second later it's in square four, which in this model effectively means that it has been blown or teleported to the far side of the universe. Even Captain James T Kirk of the Starship Enterprise might get skittish if that happened, so it's scary stuff indeed, but entirely possible and (for now) just as likely as anything else (unless Mrs Bananahead has been cheating or otherwise misbehaving of course!).

So, before we wrap up the Simple Appleverse and take a look at something with a few more apples, the Grand Appleverse, let's think about what we've learned:

- The Simple Appleverse is two apples long, two wide, and one apple high.
- Because you and your best friend, whether real or imagined and of whatever skin colour, each tossed a coin four times, the Simple Appleverse lasted for two seconds.
- The chances of an apple being in a particular space at a particular time were one in two (based on heads or tails with any inconclusive tosses where the coin landed on its side or under the sofa or in the cat's bowl discounted and retaken properly).
- Given the above, there are 256 possible Simple Appleverses.
- No one got poked in the face with a sharp stick.
- No one ate any apples while the game was in progress, at least not those apples actually involved in the game.
- Some Simple Appleverses are relaxing, while others are so scary that not even Captain James T Kirk would boldly go there in case he accidentally split his infinitives. Ouch!

I hope those few points make sense. Even Jazz is sort of nodding, although of course he may just *think* he understands, or he may just want to feel that's he's not a few fleas short of a hedgehog (hedgehogs being delightful creatures to watch but almost *always* crawling with parasites in the wild). Conversely, those of you with a penchant for reading the nerd alerts may by now be thinking, 'Aha! I see that the number of possible Appleverses is P^{XYZT},' and you would indeed be right, but we'll discuss that more in a later nerd alert, along with a little received wisdom regarding the possible values for each, but not until Jazz has gone to sit in front of the fire and lick the place where his testicles used to be.

Anyway, hurrah! It's all jolly exciting stuff. My tail's certainly having a good thrash around, and a grand tail it is even if I do say so myself. So, now let's talk about a bigger Appleverse for a bit. Yes, you guessed it: the Grand Appleverse! If you can bear with me until the end of the apples that would be great. If not, it's not the end of the world, but if you can count to over 256 it's certainly worth a go.

b)   The Grand Appleverse

So, let us assume that you're on a health kick and are about to make lots of delicious apple juice, so now you've got a big basket of apples plus an extra person with you to join in the fun … Unless, say, you're an especially reclusive mathematician … In which case you can always print off a couple of random Facebook profiles and pretend they're people you actually know … Or perhaps use a sprout named Mr Grindlevalt.

So, three 'people' in total, meaning that the Grand Appleverse will last for three seconds. We're still using just one coin, and tossing as before, so each bit of the model universe either will or will not have an apple, but now let's put the apples in a box that holds two apples lengthways, two widthways, and in which we can stack them two high, so the Grand Appleverse is very much like the Simple Appleverse but has an extra layer and lasts one second longer.

Clearly you may need to perform this experiment in (close to) zero gravity to allow you to stack an apple in the top layer on an empty space below, but then this *is* a model of the universe, and that *is* what we'd normally expect! On the other hand (or paw), some appropriately placed see-through plastic cups upon which the apples on the top layer of each box could be placed as required would work just as well for illustrating the point, and would probably be easier to achieve on your kitchen table than zero gravity.

As you can see, each layer has 2 x 2 apples, or 4 apples, and there are 2 layers, so that's 4 x 2 apples, or 8 apples in each of yours and your two friends' boxes, so 24 apples altogether. That *is* quite a few apples, but I'll wager that an astute individual such as yourself can picture the scene.

Happy? Great!

It's a good picture, because although we're about to encounter an, ahem, *slightly* larger number of possible combinations of apple arrangements, what with all those extra apples, the picture is believable, so we can use it to rationalise our subsequent discussions. And luckily, the size of the slightly larger number is of little consequence as long as you can keep the picture of the Grand Appleverse in your mind.

So, what are the chances of there not being a single apple anywhere in the Grand Appleverse during any of the three seconds of its existence? Nerdy Nomates probably already knows, but it's a 'simple' enough calculation: we just need to throw tails 24 times with no heads so, and I'll shrink the font here so that you don't think I'm just padding out the book with any old rubbish, it's a 1 in 2×2 chance.

Nerd alert!

Or 1 in 2 to the power 24, or 2^{24}, or 2^24, or in Excel

'=EXP(24*LN(2))', ie 1 in P^{XYZT} where P=2, X=2, Y=2, Z=2 and T=3.

Back to normal.

Or to put it another way, 1 in 16,777,216 which rounded down for lilting pronunciation is about 1 in 16 million, or roughly twice the number of people living in London, although of course a lot depends on who you ask and what you call 'London' (and *when*!) but the official figure from the 2001 census for the Greater London Urban Area was 8,278,251 people, so twice that is 16,556,502, so I hope you'll pardon my indulgence. Or to put it yet one more way, as a human hand was used to toss the coin, it's roughly the same as the number of human hands in London.

You can hold that thought if you like.

By the way I feel I should mention, lest you be slightly confused, that the chances of *one* box being empty are just 1 in $2 \times 2 \times 2 \times 2 \times 2 \times 2 \times 2 \times 2$ ...

Nerd alert!

... or 1 in 2 to the power 8, or 2^8, or 2^8, or in Excel '=EXP(8*LN(2))', ie 1 in P^{XYZT} where P=2, X=2, Y=2, Z=2 and T=1 ...

Back to normal.

... which for slightly different although largely the same reasons is 1 in 256, like in the Simple Appleverse, only here it would be two layers instead of two seconds, but for all three Grand Appleverse boxes to be completely empty the sum becomes 256 x 256 x 256, or 1 in 16,777,216. If you're still uneasy about discussing numbers with a dog, you're welcome to ask Nerdy Nomates, or you could get a coin and some apples and prove it to yourself, but that might take a while!

Mad-B's first random embarrassing personal secret!
Funnily enough, 16 million is roughly the number of inches between London and the Lake District where, about eighteen months ago, my humans took me and Jazz on holiday, and where Mad-B broke the decrepit and fragile metal-framed bed at 4.00 am on the first night after visiting the loo and plonking himself clumsily back down, but he begged Mrs Mad-B not to tell the cottage's owner because it was a terrible secret and Mad-B was far too embarrassed to mention it, so they used the mattress as a futon for a week, bodged the bed back together and then left. No doubt the next visitors shouldered the blame and the shame ... Unless the next visitors did the same sort of thing ... Or indeed the *previous* visitors ... And so on ... It might still be broken for all we know, but at least we can be sure that the Grand Appleverse has about 16 million different possible arrangements, and there's a tenuous link to Mad-B sitting on the bed at 4.00 am and snapping it like a twig. Hold that picture!

Back to apples.
Of course just as likely as a Grand Appleverse that's completely empty for all three seconds is one that's completely full for all three seconds: 24 heads. Analogically, that would be a bit like someone deciding to sit in their favourite armchair to read *The Word of Dog* only to find that the whole room was full of a solid lump of metal – and the rest of the universe along with it. Or possibly suddenly noticing that the entire universe was one giant great big black hole and always had been: weird stuff indeed! We'll discuss that a bit more later (unless you skip bits of this chapter), but a lump of metal will do for now. If you want to keep things simple, call it a lump of

iron. Almost everyone likes iron; you need it to make healthy blood, so what's not to like! Unless you drop it on your foot when you're *ironing*, because that usually hurts. But then you probably already know that.

Nerd 'humour' alert!

Likewise, if you don't want to think about black holes but want to be slightly more pretentious than a common old lump of iron would allow, you could think of osmium, which is almost exactly twice as dense as lead and nearly three times as dense as iron. Drop a box full of that on your foot and you probably won't be walking anywhere for a few days, except maybe just hopping to the ambulance to take you to casualty!

Back to normal.

Anyway, equally likely, although in its own little way possibly even more bizarre, would be if you threw heads eight times, then tails eight times, then heads eight times more. Then the Grand Appleverse would be a solid lump of apples one second, completely empty the next and completely full again one second after that; a bit like flicking a light switch on and off, except with apples rather than a light bulb. If I were one of the apples involved I'd be *very* confused!

Then of course, given that so far we've only looked at three of the 16,777,216 possible scenarios, there are still 16,777,213 others we could consider. But that would make this book *really* fat and boring. So we won't ... Probably not even if the publisher is desperate for a sequel ... Although that does give me an idea ... I'll come back to that a bit later ...

Naturally, some of those scenarios would involve a single apple moving serenely around the universe from one adjoining location to another, while in others there would be utter chaos that made no sense to anyone. How many do you think might make sense? There's no right or wrong answer, but feel free to play with your apples and form an opinion or theory if you like. There might well be hundreds, thousands even, but I'll wager a steak tartare to a knuckle bone that less than one in a hundred would be a nice place for an apple to live. We'll think more about that later as well. And I'll give you my email address so that we can chat about it. Let's have a quick interlude ...

Mad-B's second random embarrassing personal secret!

Mad-B can't hold his drink. He's well aware of that, but after one drink he's easily led. About fifteen years ago, for a while he was self-employed, and so 'successful' in business that he was able to pay himself a salary of at least £10 per week over and above what he would otherwise have been given in unemployment benefits. On one occasion, his accountant invited him round for a 'sympathy meal', and after a modest chilli he fed Mad-B four cans of 'Amsterdam Navigator', a deceptively drinkable lager but exceptionally strong, followed by a snifter or two of port. On leaving the accountant's house, shortly after being sick, Mad-B's legs buckled and he staggered sideways out of control, demolishing a fence panel and falling sideways into the neighbour's garden, crushing a number of the neighbour's prize-winning blooms. The accountant's wife was furious with the accountant for leading Mad-B astray. Next day she apologised to Mad-B for her husband's conduct. Mad-B's lucky like that.

Back to apples again.

Marvellous. One more section and you can skip the rest of the chapter if you like, without it being a major problem. I'd rather you didn't, but it'll be your choice. Although it probably wouldn't hurt for you to read it all. Of course if you want to quit here and now I can't stop you, but then the universe will just be a few boxes of apples and Mad-B's backside snapping a bed like a twig, and I'm not sure any of us want that.

If you do decide to read on to the end of the chapter, we're going to talk about three more scenarios: the Ping-Pongoverse, the Atomverse, and finally the Quantum String Theoriverse. I hope you believe in ping pong balls, I really do. I'm perfectly happy if you choose not to believe in atoms, although it would make my life easier if we can discuss them as if they were real, rather than just a filthy little lie they told you at school. And if you aren't saying, 'What the Bichon Frise is a Quantum String Theoriverse … I'm not sure I like the sound of that … There'd better be at least one more of Mad-B's random embarrassing secrets as well to keep me interested …' then quite frankly I'd be rather surprised, although I do have good news for you about the secret!

If by now you're wondering where I'm going with all of these model universes, please hang on in there. Basically I'm pushing your brain just out of its comfort zone, thereby leaving it open to the possibility of looking at the universe slightly differently, and therefore open to yielding to my will if that's what you decide to do. In other words, I'm messing with your head to help me achieve my own agenda, my own agenda being to help you with *your* own agenda. On the other hand, at least my agenda leads to a new vision of Utopia where no one actually needs poking with a sharp stick, and the vast majority of human angst disappears, albeit that Utopia may take some time to realise. Maybe even a few years. And some luck. So if you think that a conversation with a dog might prove too much for you, maybe you should take it easy for a bit and come back to this later. Maybe you could go and sit on a deckchair in the garden and have a nap if you like. Or tuck up warm by the fire. Whatever suits. As you may realise, it's hard for me know what the weather will be doing while you're reading this, so it would be foolhardy of me to try to tell you what to do, either in terms of where to sit *or* what to believe.

So, to summarise: I'm brainwashing you. Once your brain is washed I'm going to fill it with new information that should leave you calmer, more confident and generally happier with life. And the good news is that you're the one who gets to choose the new information. As weird cults go, you could certainly do a lot worse!

c) The Ping-Pongoverse

Hurrah! I *like* this universe!

You probably know the format by now, but this time how about we shake things up a little? We'll still have you and your two best friends, so the Ping-Pongoverse will last for three seconds just like the Grand Appleverse, but instead of apples we'll use ... yes, that's right, ping pong balls! And now the boxes will be seven ping pong balls long by seven wide and seven high. I imagine that not long after publication of this book I'll be asked to licence 'The Word of Dog Ping Pong Ball Company' (which will have be an ethical co-operative in which all workers and managers alike are paid a median wage and all profits retained to attempt to avoid redundancies in a downturn – I've never been the sort of dog who's comfy with overt capitalism or exploitation of the workers), in which case feel free to actually go out and buy them if you want to: 343 per box, so 1029 altogether, even if I can't actually promise you'll need them all, as we shall see shortly. Personally, I'd suggest that you *don't* buy that many, and certainly not just yet. If you feel you *have* to buy some, maybe just get 42 each for now. In absolutely no way do I *ever* want you to feel that you've been mis-sold ping pong balls!

Now let's shake things up a bit more. Not one but *three* coins, with all three to be tossed properly each time, and only when *all three* are heads will we deploy a ping pong ball. If that doesn't make sense, just balance three actual coins on your actual thumb and actually toss them. If they're *all* heads, give yourself a ping pong ball. One ball, not three. *Three* is the number of the *coins*, but *one* is the number of the *ball*. That's going to change the numbers, and an empty universe will be far, far more likely than a full one, but it'll be great fun!

So, what we have done now is change the chances of any particular ping pong ball being deployed to be one in eight, ie when all three coins are heads (1 in 2 x 2 x 2), and the chances of an empty space to be seven in eight, ie the opposite when one *or* two *or* three of the coins is tails. There are still only two possible final outcomes, ball or no ball, but the chance of having no ball is now significantly higher than the chance of having a ball. If you aren't confident with the maths there, please, please, please get three coins and do some actual physical coin tossing to demonstrate it to yourself, because you might just be a more visual person than a numerical one. Don't forget, you'll need to toss all three coins quite a few times to improve your chances of seeing 'typical' results, and there's always the chance that *something* weird might happen, as we're about to discover!

Right, let's have a quick look at the maths for the nerds, although if even the very thought of that makes you feel nauseous and/or tetchy, by all means skim over it and carry on with the text. The maths pales into insignificance when compared to actually picturing the textual descriptions of the exquisite beauty of the Ping-Pongoverse!

The first thing you may have noticed, mainly because we've already discussed it, is that while ...

Nerd alert!

... P is still equal to 2, because ...

Back to normal.

... there are still only two outcomes, because each space either will or will not have a ping pong ball, the chances that it *will* have a ping pong ball are now only 1 in 8, so the chances that it *won't* are 7 in 8. If you like mathematical representations of probabilities, and secretly I often do, we could therefore say that the probability of a ping pong ball being deployed in any given unit of space is 1/8 or 0.125, and the probability of it not being there is 1-(1/8), or 7/8, or 0.875.

Nerd alert!

So, the number of different *combinations* of Ping-Pongoverse is still P^{XYZT} where P=2, X=7, Y=7, Z=7 and T=3, but the probability of a completely *empty* Ping-Pongoverse for all time is now $0.875^{(343 \times 3)}$, or 0.875^{1029}, or around 1 in 4.7×10^{59}, which is a *huge* number, but we can still rationalise it, sort of, in that it's *very* roughly the number of alleged protons in 500 suns!

Back to normal.

So a completely empty Ping-Pongoverse is *really* unlikely, but *could* happen.

After that, I think I'll take a break and compose some poetry. It is a *mildly* nerdy poem, but there are no formulae and there's no rocket science. And in any case, why are we here if not to probe the boundaries of our understanding? There are *some* cats though. Not as many as on one of the pages of a much later chapter, but there are some. Insofar as one can 'own' a cat, their babies belong to Mr Schrödinger. We'll worry about who he is/was more in a later chapter as well if that's OK. For now, he's just a gentleman with a serious cat problem, both in terms of having a lot of them, and also in terms of not being sure how many of them are actually alive. I've written the poem in the shape of a Christmas tree to help assuage any concerns ...

As I was
going to St Ives,
I met a man with seven wives,
And every wife had seven sacks,
And every sack had seven cats,
And every sack had seven Schrödinger's kittens,
And every Schrödinger's kitten was
awaiting the resolution of its life/death probability,
Kittens, cats, sacks and wives,
assuming that both I, the man and his
entourage had the same intended destination,
which was now looming with immediate imminence,
and none of the wives or cats were pregnant or moribund,
how many sentient beings were likely to make it to the only
St Ives in the Ping-Pongoverse?
The end

By all means calculate an answer if you're bored. I'll show you my working in a later chapter and we can compare answers. There's no particular need to: just do it if it makes you feel cool and popular. Similarly, as promised, I'll mention a bit more about Schrödinger's kittens later for fun, as well as using it to make a bit more sense of something *really* weird, but the gist is that they're either dead or alive depending on an apparently random unrelated event, similar to tossing a coin ... I've just been informed that not everyone is familiar with the original ode from which my version was derived ... Oops ...!

As I Was Going To St Ives is a traditional English nursery rhyme that presents a riddle to which there is no logical single answer, viz (and subject to regional variations):

> As I was going to St Ives,
> I met a man with seven wives,
> And every wife had seven sacks,
> And every sack had seven cats,
> And every sack had seven kits,
> Kits, cats, sacks and wives,
> How many were going to St Ives?
> The end

Therefore, there is no mention of the direction in which the man (himself also enigmatically not mentioned in line six) and his entourage were heading, just that the protagonist was heading to St Ives, although even then the protagonist might, for example, have been heading to St Ives in Cornwall while the man and his entourage might have been heading to a different St Ives such as the one in Cambridgeshire, therefore both may feasibly have been going the right way to one of the St Ives but in completely opposite directions, which may or may not affect one's assumptions regarding the answer. Some scholars count the sacks when considering their answer to the original poem. Others do not. Each generally believes the other to be the intellectual scrapings from the very bottom of a fairly large barrel of idiots. Neither faction comments on the life expectancy, pregnancy or gestation periods of the cats or the wives, either with or without reference to the expected remaining duration of the journey. I do hope this helps.

Where were we?

Ah yes.

In the Ping-Pongoverse, the odds against a giant permanent black hole are 'big' …

Nerd alert!

… ie the probability is 0.125^{1029} or very roughly one chance in a 1 with 929 zeros after it, around 1 in 1×10^{929}. And remember, which of course you already will if you're a proper nerd, that because the little numbers are powers, 10^2 is one tenth of 10^3 and one millionth of 10^8, so 10^{929} is actually a lot …

Back to normal.

…, ie more than the number of photons (bits of light) in the universe, assuming that you take the meaning of the word 'universe' in the traditional sense and you believe in photons in the way that popular science would have us believe (ie, in very simple terms, that bits of light is what they are), which is a perfectly adequate belief at the moment, although you'll probably laugh at such folly before long when, as we shall discover, there's some extra information to consider.

'But Max,' I hear you say, 'How do you know how many alleged photons there are in the alleged universe?'

'Technically, I don't,' is the precise answer although, if you can bear with me, very soon I shall show that it's a lot less than 10^{929}. And indeed if there *were* 10^{929} photons in the universe, not that they would fit, it would be one giant big black hole that was full of white light!

We're going to be looking at ping pong balls in more depth in the next chapter, so as I mentioned earlier I very much hope that you believe in them. I'm sure you do. However, if you happen to be sitting on say a bus or a train, and the person next to you has just leant over your shoulder and said, 'I don't believe in ping pong balls,' you could suggest to them that they pop along to the ping pong ball section of the nearest sports shop and browse their wares or, failing that, maybe get them to browse the phone book under 'Therapy', ideally for someone who offers a 24-hour call out service and secure residential treatment ... Or alternatively, you could assume that they come from somewhere where there really are no ping pong balls, such as a quasi-parallel universe, but we'll deal with that calmly, simply and rationally in another chapter ...

Anyway, as mentioned, we'll think about ping pong balls more later, and I'm sure that by now you won't need me to explain the difference between the scenarios a ping pong ball might find calming or scary, viz moving serenely round its universe (box), or being blasted from one end of the box (universe) and back for every second of its existence. So I won't. By all means have a think about it though. And of course later on we might reflect on this further together. And it would be remiss and/or silly of me not to mention that while clearly ping pong balls are accustomed to travelling more quickly than apples, and better designed for suddenly changing direction, I know of no ping pong balls that feel entirely comfortable with actually being unexpectedly teleported!

d)   Atoms, Atoms Everywhere ...

Right. Let's have a quick look at the Atomverse for all of you who are still reading, which is probably most of you since I assume that you paid good money to read this and are savouring the rich experience it offers. If it's OK, I'll skip the scenarios for now, as later on I want to discuss scenarios for the universe we usually think we inhabit, but in a way that the Grand Appleverse brigade can understand, and they've already skipped to the end of the chapter. You, however, should you so choose, shall be blessed by the extra insight that a little maths can bring: I personally feel that it's usually better to sniff out beyond reasonable doubt where I am, rather than to just think I have an inkling. And we'll still have nerd alerts so you can skip over the bits of the maths that has numbers in it, so what on earth can you possibly have to worry about ...?!

Firstly, if you're prepared to believe in atoms, and you're quite at liberty not to as far as I'm concerned, you're fairly likely to believe in the simple set of particles (or wavicles for quantum aficionados) of which they're traditionally composed: protons, neutrons and electrons. You may well also be perfectly comfortable with some of the more unusual or constituent sub-atomic particles, but we don't really need them for this illustration. That said, tachyons are an old favourite of mine, as they allegedly travel faster than light, so disappear just before they start to exist. They're also essential to many branches of theoretical physics. But don't just take my word for it, let's plagiarise Wikipedia (and don't worry if you read but don't understand the following quote – no sane person does – in fact you probably need to worry if you DO understand it!).

<u>Severe</u> nerd alert!

'Tachyons frequently appear in the spectrum of permissible string states, in the sense that some states have negative mass squared, and therefore imaginary mass. If the tachyon appears as a vibrational mode of an open string, this signals an instability of the underlying ... system to which the string is attached. The system will then decay to a state of closed strings ... If the tachyon is a closed string vibrational mode, this indicates an instability in spacetime itself. Generally, it is not known what this system will decay to.'

Back to normal.

I don't know about you (if you read the nerd alert), but to me that sounds like a rather unnecessarily complicated pile of 'stuff'; but what is true by most accounts is that if tachyons don't exist, and in the 'universe' we observe they 'can't', we'd have to rewrite the whole of physics as we know it. In fact, let's do that, or at least give it a bit of spit and polish. From the babble quoted above clearly *someone* has to, so it might as well be you! With a bit of help from me. And we'll keep it simple. But you can take all the credit. Is that OK? Hurrah!

So, how big is the universe, or more importantly, how big is it going to get?

'Science', ie science before we decided to rewrite it, suggests that the universe has been expanding for about 13.7

billion years, although until everyone has read diligently to the end of this book people are likely to carry on arguing about that. If you happened to read the nerd alert in Chapter One you'll know that there are allegedly a few weird things about how big the universe has got in that time, but we'll keep it simple here: I'll exaggerate a couple of the other variable figures to compensate where necessary, so that no one, not even the nerds, should have to feel that they are being short changed.

Religion, depending on your chosen one, and not excluding the possibility that you might have validly fashioned your own religion using different parameters, generally plumps for an age for the universe of between 6,000 years in the case of Creationism, to 'it's been here forever', and while I wouldn't blame you for thinking that my claims are the jabbering ramblings of a deranged poodle, I intend to show how all three (6,000, 13.7 billion, or 'always') and more can all be essentially correct. And the funny thing is that you've already made a start on explaining it to yourself using apples and ping pong balls!

Actually, I think by definition and out of necessity *most* of the rest of the chapter is *mildly* nerdy. Sorry about that. Stick with ping pong balls if that helps. Or maybe just read the words and ignore any offending numbers? Let's share another secret ...

Mad-B's third random embarrassing personal secret!

Mad-B can't hold his drink. He's well aware of that, but after one drink he's easily led. About eighteen months ago, Mrs Mad-B invited a psychiatrist and his wife around for an evening meal. During the slap-up feast Mad-B helped the psychiatrist unwind from a stressful week, involving someone who'd been 'a bit naughty' with a Samurai sword, by joining the psychiatrist in a little wine: well over a bottle each according to Mrs Mad-B. At the end of the evening, while Mad-B avers that he has no recollection whatsoever of the alleged incident, Mrs Mad-B assures me that he stood at the foot of the bed attempting to sing Rod Stewart's 1978 smash hit *Da Ya Think I'm Sexy* while using a can of Lynx deodorant as a makeshift microphone. Even without knowledge of the alleged post dinner-party serenade, the psychiatrist's wife was furious with the psychiatrist for leading Mad-B astray. The next day she apologised to Mad-B for her husband's conduct. Mad-B's lucky like that.

Back to normal.

So, back to the question in hand, how big will the universe get? Permit me, if you will, to suggest a maximum useful age for the universe of a hundred million trillion trillion trillion trillion trillion trillion trillion trillion trillion trillion trillion trillion trillion trillion years, ie 10^{200} years. Obviously if there's a big crunch where the galaxies gravitationally attract back into one big amorphous blob that sets off another big bang, thus satisfying some of the less demanding Foreverists but not actually answering the questions of a) why it started in the first place, b) whether ultimately after the nth universe there will be an end, and c) whether the next universe will be much the same as this one. The maximum (hey, Maximum, that's *my* name!) useful age might well be a lot less than the above, but in terms of expanding, after a life of 10^{200} years, which ought to be enough even for self-regenerating non-earth humanoid aliens such as Dr Who, due to the very small but allegedly detectible (in one scenario) tendency for matter to 'decompose' to its most stable state, Iron 12, in 10^{200} years all the stars will be out and no one will know we're here. Because we won't be. Because we'll be made of iron and all the lights will be off.

Nerd alert!

Obviously a true nerd might be mildly offended by the above model for the end of the universe. Indeed I myself prefer the more avant-garde model where protons 'decay' into the degenerate matter of the predicted black dwarf stars and where protons have a half-life of just 10^{37} years, then decay further into radiation/photons, and with temperatures tending towards absolute zero bolstered only by the faint warmth of the matter decaying to energy, and ultimately with even black holes 'evaporating'. That model does, however, give a lower useful age for the universe of at best 10^{120} years, so by using 10^{200} in the earlier model we can be sure we're selecting the upper bounds of the paradigm. So, by all means feel warm and smug if you wish to, but I trust you're nerdy enough to understand the reasoning. Oh, and as we'll see, the black dwarf cum radiation model has been sullied by erroneous phrases such as 'Nothing happens … and it keeps not happening forever.' It may therefore be best to keep an open mind rather than feeling too smug!

Back to normal.

Let us simplistically further infer for now, just to get a feel for the numbers rather than to poke each other with sharp sticks if there's an instance or two when either of us would have preferred to use a slightly different number or scenario, because you're probably going to want to tweak it in a little while anyway and I'm happy to support your forthcoming desires, that the universe will have expanded from a point at the speed of light for all of those 10^{200} years (an absurd concept on several levels as we've already hinted at, and shall find out more later, but it'll do for this illustration).

Let us further infer for now (and remember that you can always rewrite this later if you like), because it says so on Wikipedia, so it must possibly be true with a bit of luck, even though it's the subject of lots of potty throwing at Physics parties, that a unit of time is roughly 5 x 10^{-44} seconds, and a unit of space is 1.6 x 10^{-35} metres in each of the usual dimensions of length, width and height.

Full scale lengthy nerd alert!

Next, because we've taken something of a guesstimate for the maximum useful age of the universe, we'll approximate the number of seconds in a year to 33⅓ million or 3⅓ x 10^7 (rather than using circa 31,558,150 for a sidereal year), and we'll round the speed of light up slightly to 300,000 kilometres per second, or 300,000,000 metres per second, or 3 x 10^8 ms^{-1}.

Therefore, the maximum useful diameter the functioning universe might ever achieve in terms of units of space (ie Plank units; the point where you just can't chop any smaller until later in this very chapter, when you can, but it's a bit weird) is 10^{200} years times 3⅓ x 10^7 seconds times 3 x 10^8 metres times 1/(1.6 x 10^{-35}) space units (which equals roughly 6 x 10^{34}), so the maximum radius, ie half the diameter, is (10 x 3 x 3⅓ x 6 x ½)$^{(200+7+8+34)}$, or 300 x 10^{249}, or 3 x 10^{251}. Therefore, taking the volume of a sphere as $4/3\pi r^3$ (yes, it's a bit of a liberty calling the universe a sphere, especially as no one has a definitive answer to the question of what shape it actually is, but we're just illustrating a point, so bite me if you must but personally I think it might turn out to be something simple like a pan-dimensional hyperbolic Möbius ouroboros, but that's just a wild guess into which no conspiracy theorists should attempt to read anything), or if we take π (Pi) as exactly 3.0 to compensate for the excess seconds in the assumed year, we simplify the equation to $4r^3$, we can see that the maximum number of useful space units at the end of the useful life of universe, where 'r' is 3×10^{251} as derived above, is $4 \times (27 \times 10^{251 \times 3})$, or roughly 10^{755}. Let's go nuts and 'round' it up to 10^{759}, ie 1,000 times as much, for anyone offended by the approximations or the lack of four-dimensional spacetime curvature distortion, ie nerds. Referring back to our discussion about photons in the Ping-Pongoverse section, we have therefore now shown that there must be less than 10^{929} photons in the universe, as even if the universe were one giant white hole there would never be more than 10^{759} at any given moment as you can't squeeze more than one 'thing' into one unit of space – with these numbers that'd be a bit like getting a box big enough to hold eight apples and trying to squeeze millions of billions of trillions of apples into it – messy!

However 10^{759} is just the number of space units at the fleeting moment at the very end of the useful life of the universe. If we want to know roughly how many space units *ever* existed we have to factor in time differently (ie instead of just using it to calculate the final size). Therefore to work out the total number of space units ever, we need to consider circa 3×10^7 seconds per year, 2×10^{43} (ie 1 divided by 5×10^{-44}) time units in a second (rather than space units and metres as we did for the size), and 10^{200} useful years. Multiply those three figures and we get a useful *duration* for the universe of circa 6×10^{250} time units, each one being conceptually the same as a box in the Ping-Pongoverse. Then, instead of dividing by two to allow for the fact it was an expanding universe (which is actually quite a surprisingly accurate way of doing it and given the approximate nature of our calculations really not worth dabbling with the integral calculus that would otherwise be involved, as we'd then have to consider the *rate* of expansion, which as we'll see later is also weird, as well as the 'sum to n terms' equation, which may offend or befuddle some readers!), we'll go nuts again and round it up to 10^{251}, and hey presto, if as with the Ping-Pongoverse before, we decide that each space unit may or may not have had, for argument's sake, a quantum energy parcel of some sort (a very small ping pong ball), the number of possible Atomverses based on P^{XYZT} becomes $2^{(space\ units)\ \times\ (time\ units)}$, or $2\wedge((10^{759}) \times (10^{251}))$, or $2\wedge 10^{1010}$, or $2\wedge$100, 000

Back to 'normal'.

As that number is far too big for any sane person to reasonably understand, plus the fact that whether a particular lump of 'stuff' exists in our universe may arguably be rather more involved than tossing a few coins, it also makes it rather pointless generating any numbers to try to explain the chances of a completely full or completely empty Atomverse, although if we were to seek to equate our universe to a game of Atomverse *they are still possible*.

Not only that, but despite the ridiculously small odds of an empty Atomverse, it's arguably still one of the more likely scenarios!

I imagine that by this stage you don't need me to tell you what's sensible, given that as suggested above we're essentially just talking about what for now we'll simply call 'the classical universe' but, technically, if you were playing a game similar to the one with the ping pong balls, anything at all in the Atomverse that sounds *good* is a possibility, as is anything *bad*. So, for example, you might decide it's 'good' if you're enjoying a piña colada under a palm tree while looking forward to your tea and some quality time with a loved one. Likewise, you may decide it's 'bad' if you're enjoying a piña colada under a palm tree while looking forward to your tea but then a horde of heavily armed flesh eating space fairies jump out of the tree and … no, let's not go there, but if you're playing a *random* game of Atomverse, anything *could* happen!

e) **Quantum Fizz**

Mild to moderate nerd alert!

OK. Last universe, and this one *is* a bit weird, so feel free to dismiss it. It is quite funky though. I rather like it. Welcome to the Quantum String Theoriverse!

It may be best if I assume that you haven't heard of String Theory, but at a quantum level it does allegedly allow for some rather exquisite mathematics, a lot of which a human named Professor Edward Witten is responsible for, with notable recent contributions to the field from Laura Mersini-Houghton among many others. Ed is so clever that even I might have trouble beating him at a game of 'Good Boy, Go Fetch the Stick' and he's widely regarded as much brighter than Einstein ... Albeit for some reason when the maths gets written down it looks rather like what happens when an Egyptian hieroglyph writer tries composing a palindromic haiku while drunk on absinthe ... Messy ... And weird ... Luckily, we are indeed blessed that we don't in any way need to discuss the messy hieroglyphs in order to fully understand *The Word of Dog*, but don't let me stop you if you wish to probe further on your own!

Anyway, the gist of string theory, and because it's a theory some people disagree with whether and how it should be represented, is that each of the traditional spatial dimensions is itself composed of a further three internal dimensions populated (or otherwise) by lots of (or some or none) quantum energy strings that are waiting for an undetermined number of conceptual Schrödinger's Coins to be tossed to make their minds up whether to exist or not. Some people prefer to think of the strings as one-dimensional because that's 'easier', but as neither model fits with anything we can *see* as such, that's probably not worth stressing ourselves about here. Ed likes lots of dimensions though. He's even got two more dimensions to mathematically unify the other nine, ie the three we see multiplied by their three internal dimensions each, into a grand unified theory of everything, and brings harmony and karma to the conflicting string theory factions under the name of M-theory. He's cool. I'd certainly let *him* rub my tummy!

Nerd taunt!

Yes, I know that's arguably a bit 'simplistic' and the phrase 'quantum energy strings' probably annoys you. Bite me!

Back to mild to moderate nerd alert.

To put it 'simply', that which seems to us to manifest itself as throbbing great big chunks of matter, such as protons, is made up of vast amounts of inner-dimensional strings of energy, and when enough collect together in one place, they start to look solid. If you'd like an analogy to help visualise a proton being made up of energy strings, it's a bit like if a ping pong ball seemed to be one of the smallest things that could possibly exist, yet the thick crusty shell is actually made up of great big clumps of energy that have made a solid, while inside every ping pong ball are billions of trillions of 'things' bouncing around essentially, for the analogy, in their own dimensions and with empty (inner) space between them, but we certainly can't *see* them.

Back to vaguely normal.

We could, if we so chose, simplistically say that in terms of energy strings a proton is a bit like a black hole in the Ping-Pongoverse, whereas an electron, which is about one two thousandth of the 'size' of a proton and behaves like a particle sometimes, a wave at others, and a rebellious freak that makes up its own rules if it feels like it, is a bit like a Ping-Pongoverse with say two ping pong balls (and we'd have to make the box 16 x 16 x 16 to get the scale about right, but it'll do as an illustration) which makes them rather more ethereal and 'ghostly', while a photon … essentially a unit of light, although 'interestingly' if you have a room full of nerds and ask them what speed light travels at, a fight will often break out because many nerds believe something subtly different to other nerds, but without fail all true nerds will always argue that light doesn't travel at exactly the speed of light ... Where was I? ... Ah yes, a photon is a bit like a Ping-Pongoverse where the boxes are actually the size of a two-storey house but have only one slightly mischievous but shiny ping pong ball hiding somewhere in each: oh, and the two-storey house would need to be roughly the size of the entire United States of America.

Nerd alert!

Including Alaska if we equate the (disputed) mass of a photon to one billion billionth of an electron volt, and we use a standard 40mm diameter ping pong ball, and the house is about 20 feet tall.

Back to vaguely normal.

What does all that mean? Well, mostly that's what the remaining chapters explain, along with what if anything we should do about it, mostly in far more simple and useful terms, along with other important questions like where Father Christmas lives, but first we'll have a really quick look, without using hieroglyphics, at the numbers that might be involved in the Quantum String Theoriverse, so please have a read even if you ignore the actual maths: at least some of the text is mildly amusing, so it won't be like the cold sweat of the nightmare of doing long division on stage in your head during school speech day in your underpants. Given that we're talking about dimensions internal to the dimensions our rationale tries to tell us we inhabit, the chances of learning an awful lot about them are currently slim (other than, somewhat ironically, perhaps through some rather weird hieroglyphical and theoretical mathematics, which I promise we probably won't mention again). What we *need* to know though, is that quantum energy strings are really very small indeed. In fact they are far, far smaller than a unit of space even, as we've discussed, when compared to a fat juicy atom. Call it the size of a squashed pea versus the Oort Cloud if that helps, ie the light year or so around the sun, 6.25 trillion miles or so, just to help you get a conceptual idea, ie if an atom was scaled up to measure 6.25 trillion miles, an energy string might conceptually be like a squashed pea. Or maybe think about a funny new sort of flat atom inside a ping pong ball. It doesn't really matter a lot because the dimensions are different anyway, although based on what we've learned so far, we can surmise the following even though we're not sure how many coins to toss:

- The chances of an EMPTY Quantum String Theoriverse are so slight that you'd need a piece of paper bigger than the universe to write down the odds in full.
- The chances of a FULL Quantum String Theoriverse are so slight that you'd need a piece of paper bigger than the universe to write down the odds in full.
- The chances of anything in between are also so slight that you'd need a piece of paper bigger than the universe to write down the odds in full.

Despite the above, the fact we're here 'talking' about it means *at least* one of the above *is actually happening*. Wow! Er, I mean woof!

What is truly magical about today, however, is that I, Max Merrybear, can tell you the answer to how many possible Quantum String Theoriverses there might be. It is ... not 42 (which in the rather splendid novel *The Hitchhiker's Guide to the Galaxy* by the late Douglas Adams was calculated over a period of 7.5 million years by a computer named 'Deep Thought' as 'The Answer to the Ultimate Question of Life, the Universe, and Everything', 'The Earth' subsequently being created to work out what the actual question was, it being *entirely coincidental*, and no conspiracy theorists should ever infer otherwise, that if you divide the 343 locations in a Ping-Pongoverse box by the chances of a ping pong ball being deployed and round down to the nearest whole number you get the 42 ping pong balls recommended for possible purchase, and 42 also being the number of questions formally answered later in this book) ...

Nerd alert!

... not (just over) 1.4×10^{51} (that's 42 factorial, ie as a mathematician would write it, '42!' ie 42 x 41 x 40 et cetera all the way to x 3 x 2 x 1 ... it's a maths joke ... sorry ... but then if you're a nerd you probably liked it ...

Back to mild to moderate nerd alert.

... The answer is (cue drum roll): two to the power of a googolplex!

If you aren't sure what a googolplex is, it's ten to the power of a googol, which sounds simple but is actually quite hard to imagine, as a googol is ten to the power of a hundred, so a googolplex is 10^10^100. It looks like it'd be easy to write out in full, but it's actually yet another case where the piece of paper would allegedly be bigger than the universe. 10^2 is a hundred, 10^3 is a thousand, 10^12 is a trillion, and 10^1000000 ... etc ... 000000 ... etc ... 000000 is BIG.

Back to almost normal.

A googolplex does, however, pale into insignificance

compared to the sexy number *Two To The Power Of A Googolplex*.

Ah, the beauty of coming up with such a stupid number, and then surmising as if by magic that it's the number of possible different Quantum String Theoriverses. *But it could genuinely be true.* And as we'll find out later, I've had a bit of help in surmising that number. And it's not as stupid as infinity, which we have already, perhaps without realising it, now proved cannot exist in this universe, in that 'infinity' is bigger than the product of the maximum total number of space units and/or energy strings multiplied by the maximum total number of time units. That's right: infinity is now *wrong*. Whatever any teacher, scientist or mathematician tells you or has ever told you to the contrary, and I am actually being perfectly serious here, anything with 'infinity' in the answer is **wrong**. On that subject, as we shall see later, some of the finest minds already agree with me. But today, for the first time, I hereby affirm that the biggest possible answer is now $2^{GOOGOLPLEX}$.

BIG is the new infinity. Infinity or any other theoretical numbers larger than $2^{GOOGOLPLEX}$ are just glib generalisations or errors used by people who have not yet read *The Word of Dog*.

Back to completely normal.

Let's talk about playing god.

Chapter Three

Kevin and Sarah's Own Little World

Permit me to assume, if you will, that you have at least read and understood about the Ping-Pongoverse given that, as long as you have the bloody-minded determination and a couple of 'friends', you can actually physically demonstrate that particular universe in action to yourself on your own kitchen table. Paw on heart, I assure you that I firmly believe it's worth trying to understand. If you're not sure about it, please, I implore you, go back to *Chapter Two, Section C*, and study it diligently until you understand the concept of ping pong balls bouncing around inside a cardboard box, and the basic set of rules that they obey in the game, ie whether or not to 'exist' from 'second' to 'second' based on tossing a set of three coins. Either that, or wherever I've written 'ping pong ball' you'll just have to read 'apple', for 'Ping-Pongoverse' you'll just have to read 'Appleverse', and wherever anything to do with something so small that the eye cannot see it unaided, for want of a better alternative, you may have to read 'apple pip' and just pretend that your eyesight's bad and the apple pip is so tiny that you need a really big magnifying glass to be able to see it. It is *your* life and *your* choice of course, but your life may nevertheless be more rich and fulfilled if you're able to think about ping pong balls and the rules they obey in the Ping-Pongoverse.

But what if you decide to bend the rules?

What if you decide to play, at least with respect to the Ping-Pongoverse, god, or indeed the name of any other particular deity you decide to choose (or leader or demi-god of a group of deities, eg Odin or Mars), whether or not you believe that deity exists in real life; therefore, as far as the Ping-Pongoverse is concerned, you personally are their god, whatever the nature of the rest of the universe. So for argument's sake in this *game* you are a god. After all, it's your house and your ping pong balls, so your friends probably won't mind having another game, this time playing by any 'House Rules' that you decide to make up, especially if one of them is a carrot. Otherwise, if they refuse to have a game based on your house rules, they might not feel welcome in future, or perhaps you might cast one or both of them out and invite new chums to play next time. Or indeed you might boil and eat them.

So, for the rest of this chapter, *you* can be the benign or malignant ruler of all you survey, at least with regard to ping pong balls and indeed, just for now at least, solely your own personal private ping pong balls in your own personal private (or shared) kitchen (or something you can pretend is a kitchen), rather than all ping pong balls everywhere. Please do not covet thy neighbour's ping pong balls. That would be *rude*.

Leaving any mischievously malignant urges for later, what rules might you decide upon to try to make your personal Ping-Pongoverse more aesthetically pleasing: the sort of Ping-Pongoverse where a ping pong ball might be happy to live? Here are a couple of thoughts for you to ponder, although you are of course entirely welcome to make up some of your own rules and philosophise about the consequences of those instead.

Firstly, you might decide, and the effects would be dramatic, that no ping pong ball would be deployed in a section when doing so would mean that it touched another ping pong ball.

Have a quick think about that. Rule one, for this game only: 'No balls touching.'

It's entirely up to you whether you want to picture this in three dimensions. If not, we can go along with the simple model of just two dimensions, ie one layer of the box, so a square of seven by seven (yes, technically that's three dimensions anyway, with the third being just one ping pong ball high, so excuse the brevity of my terminology if you will: remember, not *everyone* who reads this will be as astute and observant as you, ie the person *currently* reading this!). Therefore, for example, if you start by tossing head, head, head, you would deploy a smiley ping pong ball in section 1 thus:

☺	2	3	4	5	6	7
8	9	10	11	12	13	14
15	16	17	18	19	20	21
22	23	24	25	26	27	28
29	30	31	32	33	34	35
36	37	38	39	40	41	42
43	44	45	46	47	48	49

Ping-Pongoverse

Layer One

Squares 1 to 49

Incidentally, by the preceding choice of phrase I do not wish to imply that you should actually draw smiley faces on your ping pong balls, and indeed it would assist greatly if you can keep at least one of them with nothing drawn on it for a later chapter, although by all means draw on some of them if you like: after all, they are *your* ping pong balls!

This then means that you don't need to toss for sections 2, 8 or 9, as if you deployed a ping pong ball in any of those sections it would mean the new ball would be touching the ping pong ball in section 1, which is against the 'no balls touching' rule you just made up. Of course if you so chose, you *could* decide to take the goes for sections 2, 8 and 9 anyway, then afterwards see if the result obeyed the rules, then start again from scratch if not. By all means do that later for fun if you wish, although if it's OK with you I'll use the simpler method for now!

Mild nerd alert.

If you're thinking in three dimensions (and it's absolutely fine if you aren't, because it probably means you are more normal and well adjusted than a true nerd) this also rules out sections 50, 51, 57 and 58. And away from the edges in three dimensions, a single ping pong ball would need all of the surrounding 26 sections to be blank.

Back to normal.

I've just done the experiment myself and it looked a bit like this (and I promise this was absolutely genuine, albeit it was the only time I've got something slightly weird to happen, and if you do actually have nerdy tendencies you might well happen to notice and/or calculate that it's not really *that* unlikely anyway):

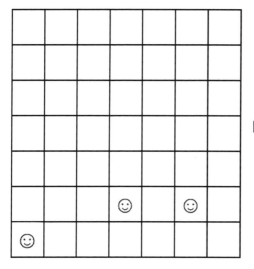

Max's Crazy Ping-Pongoverse!

But the 'no touching' rule means that your balls might feel lonely. Assuming that you're still feeling benign rather than vengeful, that's not ideal for you or your balls. Mad-B had my balls chopped off when I was six years old and had something called testicular cancer, but wherever my balls are now, I hope they aren't lonely! ... Sorry ... That was probably inappropriate ... When you're a dog you kind of get used to castration jokes ...

So, let's see what happens if you decide to make a rule that whenever a ball is deployed, before we've *finished* arranging the Ping-Pongoverse it *has* to be touching, for argument's sake, one but *only* one other ball. That makes it a bit more complicated, but nothing too hideous to contemplate. In the easy case where you just use one layer, so it's a seven by seven grid rather than worrying about the whole three-dimensional box, and where you still toss all heads on the first go, so you deploy in section 1, what it now means is that either section

2, 8 or 9 *has* to contain a ball so that the first ball doesn't get lonely. Of course that could mean they're touching horizontally, vertically or diagonally, so by all means mess around with the rules at your leisure if you think balls touching diagonally is a perversion in the eyes of the lord of the Ping-Pongoverse ie, in this case, yourself.

Therefore, for example, if the throws for sections 2 and 8 are 'negative' (although naturally no one is telling you that you have to address the sections in any particular order; I'm just taking the obvious option here, or at least the option that seems obvious to me – feel free to make up your own rule about the order in which sections/squares are considered if you like), ie they each have at least one tail, you *have* to deploy in section 9 (because it's your game and your rules) or the ball in section 1 would be lonely, so in that case you don't actually need to bother tossing for section 9 because you've already decreed the result by divine application of the will of the lord. And let us not forget that thou art mighty and might otherwise smite us down in thy displeasure at us, for thou might be benign for now, but might decideth to telleth us not to tryeth thy patience!

Similarly, if the result for section 2 was positive, ie all heads, the ball in section 1 would acquire its partner straight away, so you wouldn't need to toss for sections 3, 8, 9 or 10 as no balls would then be allowed in those sections, or a threesome would occur.

I *could* at this point go in for some fairly hard core padding of the book by looking at various scenarios in the full three-dimensional Ping-Pongoverse and drawing, then discussing, each scenario using seven grids for the seven layers, but it wouldn't really add any value and I think it might alienate many of the lovely people with whom I most hope to connect, so I shan't. So if the book seems a *bit* thin for whatever cover price it ends up being sold for, remember that's because I *love* you. And to be honest, if you're nerdy enough to want a full three-dimensional model, I'm almost certain you should have the wherewithal to be able to build or picture it for yourself. If you want to actually physically build such a box for each of you and your two chums, I recommend using some small transparent Perspex cubes (similar to the transparent plastic cups of the Grand Appleverse) deployed where ping pong balls are absent so that a coherent three-dimensional model is possible; I've recently sent Mad-B off muttering to the garage to build just over a thousand Perspex cubes for me so I can actually build such a model, although I'm not sure he thinks it's a good idea. However, if perhaps you're just a bit nerd-curious you might choose to represent it on paper with seven grids each of seven squares by seven, with Grid One numbered 1 to 49, Grid Two numbered 50 to 98 and so on to Grid Seven numbered 295 to 343, then repeated twice more to show all three seconds. I can't promise it'll be as much fun though, but at least there are various options!

So: two dimensions and just imagining it are fine. That will tell us all we need to know. It's a simple seven by seven grid for each 'player'. Job done. If I'm honest, the Perspex cubes are really more something to keep Mad-B occupied for a bit. He's a *terrible* fidget when he's bored.

Of course if you decided to change the rules again so that any given ball had to touch *at least* one other ball (rather than *only* one) you may well get different patterns of polygamous balls, although it might well mean a bit more coin tossing on your part, as you couldn't then preclude a section just because two balls were already happily joined nearby. That said, in many ways I quite like the idea of two balls joining and then staying together as a couple. A canine psychologist or dog whisperer would probably say that I'm compensating for my disappointment at my parents, as they only met once ... And money was exchanged ... Still, I guess I'm not the only one with parental issues.

And what if you decided to keep changing the rules, say so that if three balls were touching they had to have two clear spaces, and three spaces around four balls touching? Or what if you decided to increase or decrease the number of coins tossed depending on the proximity or otherwise of other balls? You could probably keep thinking of new ways to play the game until your head hurt, but that's OK because it's *your* Ping-Pongoverse!

And what about when your chums take their goes? Would you want your exquisite patterns suddenly strewn to the four/eight corners of the oddly square/cubic Ping-Pongoverse? You can if you like: you're the one in charge of making the house rules. Or to avoid the chaos you could make some rules to say how your balls behaved from one second to the next.

Let's go back to picturing the simplest one-layer no-balls-touching Ping-Pongoverse for a moment so that we can think about 'time', ie each of the three of you having a go, each with a one-layer Ping-Pongoverse, each representing one second of time.

Let's pretend you have first go and, and it's unlikely but entirely possible, that you just deploy a ping pong ball in section 1. Your best chum, Chum-1, has the next go and just deploys in section 25 (the square right in the middle), and then Chum-2 has a go and just deploys a ball in section/square 1, exactly the same as you did.

That sounds like a *horrible* universe!

No, let me rephrase that. *I* think that sounds like a horrible universe, and if you like it I might think you're a bit 'out there', but I certainly wouldn't try to bite you let alone *kill* you for liking it. In fact Picasso would probably *adore* it if he weren't dead unless, for argument's sake, Picasso's ghost 'lives' in a bucket of salt water on your kitchen table, although if it's OK with you I shall wait until much later before contemplating the possible reasoning for you having such a bucket positioned accordingly and the likelihood or otherwise of it being haunted by Picasso.

And I might even decide to defend your *right* to like that Ping-Pongoverse, just as long as you don't want me to hang a picture of it on my lounge wall above the fireplace and look at it admiringly every day, or to have a gravity-defying three-dimensional hologramatic projection of it hovering above my mantelpiece if you're still thinking in three dimensions. And yes, I am a dog, but I do live in a proper house with a proper mantelpiece and a proper fireplace, thank you very much!

Anyway, what worries me about that universe is that your population of sentient beings, aka in this case the solitary ping pong ball, starts life at the edge of the universe, then the next second is blasted into the centre of the universe, albeit seemingly intact if perhaps somewhat befuddled, then the following second is blasted right back to the edge of the universe again. If I lived in your universe, I'd certainly feel rather skittish!

So, house rules. And if you're playing the part of a benign omnipresence you might decide to change the rules so that things work out better for your ping pong ball(s). And I think I'll just use the word 'square' from now on because 'section', although perhaps more technically appropriate, sounds a bit like 'second' when I hear humans say it, so it might otherwise be confusing. They're 'woof' and 'woöf' in Canine, so completely different!

Assuming that you still take first go, let's see what happens when you apply the following rule(s): 'During each second of the Ping-Pongoverse, no ball shall be deployed unless in the previous second a ball existed within one square in any direction of the square currently being considered by the tossing of coins as needed, and no matter shall be created or destroyed after the end of the first second of the Ping-Pongoverse.'

What happens now, assuming that in the first second of the Ping-Pongoverse we just create a ping pong ball in square 1 as before, is that next second, ie in Chum-1's grid, according to the rules, we *have* to create a ping pong ball in either square 1, 2, 8 or 9 (to avoid matter being destroyed and needing to keep the matter within one square of square 1), but *can't* in any other square (to avoid matter being created after the first second). Then in the third second, ie Chum-2's go, if say the ping pong ball has moved from square 1 to square 9 during Chum-1's go, then in the third second it might appear in any of the squares in the ping pong galaxy below, but the rest of the universe would be off-limits!

1	2	3
8	9	10
15	16	17

I like that Ping-Pongoverse. The ping pong ball moves around a bit, but in an orderly manner and with no particular traumas, apart perhaps from feeling a bit lonely because of the simplicity of our example. Perhaps, if you're feeling benign, you might like to repeat the experiment, only this time using the 'two balls must be touching' rule. Or you could just imagine it. OK? Let's do that. Close your eyes and think for a moment ... Ouch!

Did you see what I saw (which quite frankly is relatively unlikely, but stranger things have certainly happened)? My ping pong balls obeyed the rules, but in the first second they were in squares 1 and 2, the next second it was 2 and 10, then finally it was 16 and 17. That sounds *painful*! Here's a picture:

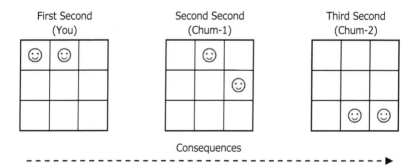

Consequences

What actually happened there was that the ping pong balls started touching horizontally, then the next second it was diagonally, then in the third second it was back to horizontal again. I'm not sure I like all that wriggling around. If I were looking to improve my house rules I think I'd want a bit more stability, yet still allowing occasional zany or amusing motion, so I might amend the rules to specify the minimum time that had to elapse between two shape changes. One thing's for sure, running a single layer Ping-Pongoverse isn't as easy as it sounds! And can you imagine the rules if we were looking at the full seven layer Ping-Pongoverse? Well, the good news is you don't have to if you don't want to: as already mentioned, one layer is plenty for illustrating our examples, but please don't let me stop you if you fancy having a third dimension in your own private personal universe.

So. Running a Ping-Pongoverse. Not easy.

And we haven't finished yet!

What if you took your turns with no rules at all, like we did earlier, and afterwards you surveyed your Ping-Pongoverse through time and didst think unto thyself, 'Lord, I see that mine own Ping-Pongoverse is good, yet I dost greatly admire the Ping-Pongoverse of Chum-1 with his central ping pong ball. I thinkest unto mineself that I might just messeth about with time for a while.'

And so you, god of the Ping-Pongoverse, might decide to apply the rules of motion starting from the hallowed point of the central ball of Chum-1, ie the ball originally in square 25, and moving back through time to retake your own go, for thou art mighty and can do that. Of course you'd probably also make Chum-2 retake their go, but that's just rewriting the future. Any idiot can do that. It's changing the past that only someone as mighty as yourself can orchestrate!

Let's close our eyes for a minute and imagine that: you retake your go, but you *have* to deploy a ping pong ball and it *has* to be within one square of square 25. You also make Chum-2 retake their go, also deploying within one square of square 25.

So, you changed the past and the future, both according to some rules you liked the sound of, based on somewhen in the middle of 'time' that pleased you. Motion in your Ping-Pongoverse *might* now, throughout its history, with the first and third seconds revised based on the 'golden era' of the second unit of time, look like this, with the middle second influencing (but not completely dictating) what happened in the past and future, but with the overall consequences looking like a perfectly reasonable sequence of events from beginning to end:

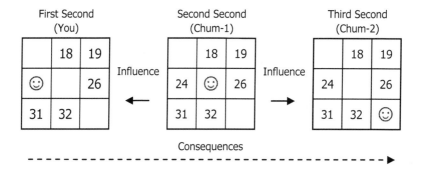

Does that make at least a reasonable degree of sense, even if you wouldn't yet want to stand up in front of a crowded room and give a presentation on it while wearing just yesterday's undergarments? If not, I'd suggest reading it again and maybe drawing a few diagrams of your own Ping-Pongoverse (if you actually have one of your own, rather than just having decided to go along with mine for now, which let's face it is what many people will probably have done!). It's certainly worth the effort to understand that you, as ruler of the Ping-Pongoverse, while playing this game at least, can move through time (and fiddle with the odds by making different rules for tossing coins as well if you really want to) and change things for what you hope is the better.

Of course it's roughly equally as likely (although influenced by the effects of considering each square sequentially or otherwise), based on tossing coins, that your attempt to change the past and future might mean that the ping pong ball just got recreated in square 25 for your go, and just stayed there for the go of Chum-2. That would make for a fairly boring Ping-Pongoverse. A kind and benevolent ping pong deity, and I think I shall, if it's OK with you and to make you sound more omnipotent, refer to you for now as 'Pingpongor', might decide to amend the rules of their universe to make motion more likely, perhaps by specifying a time after which a ping pong ball is *no longer allowed* to stay in the same square, so perhaps, as a rule subservient to the other rules your Ping-Pongoverse is currently obeying, you might perhaps decree that: 'Whensoever a ping pong ball hast not moveth for two seconds, on the third second shall it always move.'

You could also optionally tag on: 'Lestwise I shall smite it down and it shall exist no more, for I, Pingpongor, am a mighty lord and not easily pleased by slothful balls.'

With both the above rule and the suffix (ie the smiting

rule) in place, what can and eventually almost certainly would happen, especially if you get a lot more chums round to play a longer game, is that you smite all of your balls and your universe disappears. That may or may not fit your plan, but it would, if left unchanged, be 'game over', which is fine if you want the game to end, perhaps once you've travelled back and forth through time to assuage your mighty will until the Ping-Pongoverse pleases you greatly, so much so that you've even kept a diagram or home video of exactly how it functioned over time (in all of its layers if you're playing in three dimensions). Perhaps first thing tomorrow evening, after your day's toils, you might decide to visit your local Ping-Pongoverse Club to enter it into the regional Ping-Pongoverse championships, confident that you, Pingpongor, shall emerge victorious through the regional, national and international competitions until ye be crowned The King of All Time, or something equally as grand. You might even, as the winner's privilege, get to choose your own title. Kevin the Awesome, for example. Or Sarah the Omnipotent. Whatever you like.

Or maybe you don't want to smite your balls.

Maybe, to create a bit of movement, you might decide to make the rule that (and you might want to read this bit slowly): 'Whensoever a ping pong ball hast not moveth for two seconds, on the third second, whether I currently decree that time moveth forth or back, a new ping pong ball shall be created in the lowest numbered adjoining empty square if one be available.'

That sounds like fun. Woofhoo!

There are, of course, two new and slightly different problems now, especially if a lot of chums turn up unexpectedly like before, such as when you win the lottery or maybe get a large advance from a respected publisher for being the first literary puppy to help a wide audience solve all the mysteries of the universe, and the new chums all want to take turns, so your universe carries on for far longer than three seconds.

Firstly, as I'm sure you'll have no trouble imagining, if you keep creating matter, sorry, ping pong balls, in a box, if you don't have a smiting rule sooner or later it's probably going to fill up. Then it becomes a black hole (in this game), given that your balls are mashed in as tightly as possible with no hope of ever having a good stretch, scratch, or indeed a bounce around. After the initial excitement of thinking, 'Woohoo, I've created a black hole the size of the universe,' when you realise a black hole is all that's going to happen now and for evermore, it becomes a bit dull (because there's nowhere left for anything else to happen in, and with the *current* rules, ie 'whether I currently decree that time moveth *forth or back*,' if you create matter in the past it'll also need to be populated forward into the future, which gets messy).

Alternatively, if you don't want your ping pong balls smote or mashed into a black hole, the only option when you're adding to the number of ping pong balls inside the box (from your stash outside the box) is to expand the size of the Ping-Pongoverse occasionally. Luckily, that's easy: you just get a bigger box. If we were trying to build a Ping-Pongoverse in a balloon it would still be easy (ie to expand it would be easy – the bit about keeping your balls in the right place might be a tad more awkward if your universe was balloon-shaped and wobbling), in that a spot of divine wind from the breath of you, Pingpongor, would do it (ie you could blow the balloon up some more by blowing into it, just like blowing up a balloon, which is exactly what you would be doing, although you might first want to do a quick risk assessment concerning the possibility of instantaneously destroying the universe by popping the balloon), but we aren't ... So we don't need to think about it ... So it would probably be easiest if you don't ... We can think about it conceptually together later on if you like ... For now, the Ping-Pongoverse is just a box (or a square) ... If you want to expand it you just need to get bigger box (or draw a bigger square) ... That's it.

BUT.

To avoid worrying about expanding the universe in the past and then worrying further about how to handle the present and future, what you might well decide to do is only expand the Ping-Pongoverse whensoever you goeth forth through time. You don't have to, but you might, so for now that's what we'll do. So your divine rule might become (and you might want to read this bit even more slowly, although a simplified version will follow): 'Whensoever a ping pong ball hast not moveth for two seconds, on the third second, when I moveth forth through time, a new ping pong ball shall be created in the lowest numbered adjoining empty square if one be available and, once time shall have proceeded forwards according to my will, in the next second embiggened shall be the Ping-Pongoverse, and "one" shall be the number added to each spatial dimension for every second during which through divine creation shall have been made a new ping pong ball.'

Or to put it another more slapdash and imprecise but legible way, although perfectly valid for the purposes of the explanation, 'Whenever I create new matter by adding a ping pong ball, to make sure there's still room I'll make the box a bit bigger as well.'

The bottom line is that there's a world of fun to be had with a Ping-Pongoverse. You might like to think of other rules and see how the model behaves. Or you might just like to carry on reading. It's up to you. You could make the expansion (bigger box), if triggered, dependent on the separate toss of one or more coins, perhaps to limit the rate at which your Ping-Pongoverse expands, ie perhaps only get a bigger box on average every eight times you create new 'matter'. Or you could, for example, throw say ten coins for each existing ping pong ball and if they're all heads (a 1 in 1,024 chance), smite it from existence to see how that balances the matter being created. Matter can therefore be both created *and* destroyed based on a combination of your divine will and a bit of random luck.

Oh, and by the way, if you're thinking 'but surely matter/energy can neither be created nor destroyed', I'm afraid you were probably lied to at school, however unintentionally, possibly to get you through some sort of science exam where the First Law of Thermodynamics was appropriate to the question in hand. We'll come back to creating matter later, when we discuss how matter gets created, which it does, hence there being a universe that has something weird going on in it ... Maybe keep an open mind for now ...?

And that's nearly it for this game of Ping-Pongoverse.

Except for one thing.

Gravity.

You may well have played, seen or heard of the game 'Table Tennis'. I've never played it myself, what with the lack of opposable thumbs, but I have watched it being played and have noted that, like so many other things, the table tennis balls generally fall towards the table due to the gravitational pull of a somewhat larger 'ball', ie the earth. If you're feeling really clever or nerdish, you might like to make up a rule for your Ping-Pongoverse to encourage balls or clusters of balls to feel more inclined to move towards each other rather than to behave more randomly or to move apart. Or you might not (because it would be *hard*!). You might want the opposite, so that if the Ping-Pongoverse expands according to your divine will, the newly formed outer regions might more quickly be populated with balls rather than existing largely as an empty, motionless void. Or if you feel super clever, you might want to make up a rule that would allow ping pong balls to attract whilst still populating the outer reaches of your universe, just like in some of the popular explanations for the universe we live in. Ouch! That makes my head hurt! How about you?

And we haven't even worried about keeping the balls the right way up yet. Although of course if you drew smiley faces on some of them earlier you may perhaps have already thought about it, but most of the rest of us probably aren't going to worry about it with you, at least not too much. Sorry, but by all means have fun though.

Of course what we've done with the Ping-Pongoverse, we can fairly easily scale up to apply to the Atomverse. The only real difference is that it's rather more awkward getting the Atomverse to fit on your kitchen table. I'd take the Atomverse outside if I were you. There's much more room outside. In fact the sky's the limit. Literally!

And the Quantum String Theoriverse is really not that

different, apart from the whole thing with the internal dimensions that allow more exquisite maths at a quantum level, or in our case, and feel free to disagree, among other things allow the spatial dimensions we see to behave like solids we can touch, while on a smaller scale they can be a lot less well defined and behave like waves that do and don't exist in many places simultaneously … Er … apologies if you skipped the end of the last chapter because it all got too scary … Um … maybe it's worth going back to have a look at *Chapter Two, Sections D and E* after all unless perhaps, in the simplest case, you're still happy just calling everything an apple or an apple pip!

Of course I've known that all along, because I've edited and re-edited this book a dozen times in minute detail, just to make sure the translation from Canine to Human is as perfect as I can make it, so I humbly beg your forgiveness for taking many of you through a gradual iterative process to make you increasingly comfortable with the information contained herein: I can say paw on heart that my intentions were honourable.

Anyway, finally, just out of interest, now we've cleared the air and can progress in a new spirit of openness, here is a picture of what might *actually happen* to a galaxy in a game of Quantum String Theoriverse. Of course it looks exactly like what might happen in a game of Atomverse, because humans and dogs can only 'see' large objects, such as atoms, or perhaps some of the more noteworthy sub-atomic particles. And it is on the face of it *most* unlikely to *actually* happen, but in terms of an understated tenuous link to a future chapter or two, quite frankly it's awesome!

However, for those of you still doggedly thinking about apples, here's a picture that should at least challenge you philosophically ... Someone who just read a pre-publication version of this just asked me if I was trying to say that God is a table tennis bat. What do you think? If you *also* think that was my intention, I can only apologise for the confusion, although I really do hope it was a one-off ...

Right. That's it for this chapter. Now we've had a thorough grounding in the life of ping pong balls, it's time for you to meet Pyror!

Chapter Four

It's a Dog's Afterlife

Do you like fireworks? I do. Mad-B used to let some off on the patio every year when I was younger, and I'd watch fixatedly through the (closed) patio doors with my previous Labrador chum, Shadow. A few years ago she also died, although unlike me she didn't get better. I do sometimes miss her still; she was like an adoptive mum to me when I was little, and it was before Mad-B met Mrs Mad-B, so Shadow was like my first adoptive mum. But she's OK. Or at least she will be (which may sound like an odd use of the future tense for now, but please bear with me), and eventually she'll be happy for ever. I hope I'll see her again, but things will be different for sure, even if it's just that her kidneys will be working better next time. I expect that one or two other things will be different as well. And you've probably got chums, human or otherwise, that you'd like to see again one day (and probably some who you wouldn't), although there's a fairly good chance you would also like there to be one or two changes, even if they were just changes to yourself, like a leaner stomach or, in my case, a full set of testicles. I dare say that for many of you there may be a few other changes you might like to make at the same time, like your boss always being on holiday, or a few more creature comforts for a few less hours a week of thankless toil, or someone you love being happier than they are.

In Shadow's case, she'd probably wish that after the weekly bath she had to mitigate her incontinency smells I didn't hump her so enthusiastically or insistently, but then we all have our little foibles or our own little crosses to bear, and I was young and less empathetic than I am today.

But back to fireworks.

Whereas I love them, to Jazz they're the apocalypse. As soon as the banging starts anywhere within a mile or so, he tries to find somewhere to hide, although nowhere feels safe, and he cowers and trembles, waiting for the anticipated slow and painful death. I guess that's got something to do with our respective upbringings. Mad-B has always given me lots of cuddles and often calls me his Special Little Gentleman, although Mrs Mad-B did raise a furrowed brow when she first heard Mad-B call me his Furry Superman, so from my perspective if Mad-B let off colourful fireworks it was bound to be fine. As I mentioned earlier, Jazz was a rescue dog who'd been starved and tortured, and while Jazz can never bring himself to talk about it, we can tell from the whens and wherefores of his cowering that it was the male head of the household who did the torturing. Jazz likes and loves Mad-B well enough; it's just that when he has flashbacks, even though Mad-B doesn't let off fireworks now, Jazz gets so scared he just collapses and waits for the end. And although deep down I feel suitably contrite for thinking this way, I confess that it is rather funny, and also not uncommon, when Mad-B is following Jazz along the hallway, for Jazz to just drop cowering to the floor in front of Mad-B and make him stumble or even trip over. Mad-B has a special phrase he uses when that happens. I'm not 100% sure what it means, but he doesn't use it when we have visitors apart, occasionally, from referring to Jazz by the word 'Sméagol' (pronounced 'Smeegle' and who, I understand, is an undesirable character in a literary work and film series called *The Lord of the Rings*).

And the main reason Jazz is terrified of fireworks is that because, like many a dog, he believes the universe is essentially a giant firework, and the fireworks humans let off are a flashback to the primal id of creation itself, ie fireworks often make big bangs with lots of hot fizzing and sparks that emulate tiny galaxies. And whereas for me creation was fun, and so fireworks are fun, for Jazz, for the first year or so, a lot of his creation meant primal horror: he grew up in what from his perspective was a hell he could neither endure nor, for that first year, escape.

So. Canine religion. And why should you pay any attention to it even if some of us are furry superdogs? The simple answer is, of course, that you shouldn't. You're welcome to if you'd like to, but no one's going to bite you if you don't. Well, no *good* dog would bite you for it. As you probably already know, some dogs are stroppy little so-and-sos just like some humans, especially the ones with extreme runt complexes, like Adolf Hitler and many Chihuahuas. In fact if I were you I'd just have a *think* about canine religion for now and see if it makes any sense to you as it unfolds. If it does, by all means borrow a few bits, most, or even all of it, and use it as a basis for your own system of beliefs. As we shall see, that's the good thing about canine religion: it's surprisingly adaptable.

So. The universe is a firework, probably some sort of big impressive rocket, rather than a Catherine Wheel, although Mad-B says that some days it feels more like a Jumping Jack, which he tells me in the 'old days' were a bit like a Chinese firecracker that you threw on the ground. Whenever he mentions them, he also tends to have a rant about some 'faceless do-gooders' doing risk assessments and getting Jumping Jacks banned on health and safety grounds. It can get confusing listening to Mad-B.

Please don't worry where this is heading yet!

Anyway, the chances are that the rocket/universe was set off in some sort of explosion *for some sort of reason* to be discussed more later, as mentioned in a sort of cosmic big bang, or so goes the simple version of canine theology. And the conceptual firework was designed by Pyror, who then lit the conceptual fuse that subsequently conceptually launched it, and because Pyror designed the firework, in effect it's behaving according to his house rules, very much like when you were playing Ping-Pongoverse, only on a slightly grander scale. And if it's OK with you, I'll refer to Pyror as if he were real for now. It's up to you how you interpret that, and indeed whether he has two legs or four and whether he actually has a gender and if so what if anything he does about it.

Please don't worry where this is heading yet!

Of course if at this point I were to glibly claim that the universe is actually just a firework designed and launched by some sort of deity, I think you would be well within your rights to label all dogs everywhere as stoopid, and to go straight to the nearest bookshop and demand a refund; not necessarily the retailer you bought this book from, whether online or offline, just the *nearest* one.

Luckily, it's not that simple, and indeed far from it, especially if we're going to demonstrate how everyone, scientists, theologians, Christians, Jews, Muslims, Jedi, Shamans, cats and dogs can be right at the same time, let alone what it all means for everyday living, so please don't think this is all we're going to discover or learn about 'the great questions', because the opposite is very much true!

You already know the beginning of the basis for starting to build a comprehensive and intelligent answer of course.

It's $2^{\text{GOOGOLPLEX}}$ and the Ping-Pongoverse ... Actually that's a bit simplistically glib as well ... Let's take it one very small step at a time ...!

Please don't worry where this is heading yet!

Firstly, exactly when did the universe start? Let's look at some diametrically opposed opinions from two disparate factions, a small subset of whom might sometimes allegedly enjoy screaming 'Cretins! Die!' at each other.

In Creationism, not that this forms the basis for canine religion unless of course a dog chooses to believe it, it was about six thousand years ago, but then what do we mean by 'start'? When the fuse was lit, when time began, or when *matter* began? If we're considering $2^{\text{GOOGOLPLEX}}$ possible universes, ie all *possible* universes with no house rules, time could have started long, long before there was any matter there to fill any of the space, and there's no reason why an empty void couldn't have expanded emptily as time progressed. In terms of the probability of it happening with no house rules (other than perhaps to cover the expansion) it's just as likely as a whole lot of other possibilities, and a whole lot more likely than many others. Each possible universe on its own would be so unlikely that it might seem too utterly ridiculous even to contemplate, but just like if you sold a million gazillion squazillion tickets for a raffle, *one* of them would *have* to win, or in this case *happen*.

Please don't worry where this is heading yet!

A Canine Creationist might therefore argue, and I'm on sniffing terms with at least one human Creationist, that in terms which avant-garde science may find more palatable, the universe was created by Pyror as a firework 13,700,000,000 years or so ago, but that matter, or at least significant matter, such as that which one could say existed beyond the level of quantum uncertainly, didn't occur for the first 13,699,994,000 years of that. In an unbridled no rules game of Quantum String Theoriverse *this is actually feasible*. You could of course demonstrate this phenomenon to yourself in a reasonably undemanding timeframe using, say, the Grand Appleverse model and two chums where in seconds one and two no apples are deployed, yet in second three the universe is populated by a dazzling array of exquisitely arranged apples. I imagine though that as god of the Ping-Pongoverse, rather than using apples you might prefer to demonstrate it using a Ping-Pongoverse with no house rules, although of course with so many squares, ping pong balls and perhaps even the full seven layers, it might take a while!

So, while our minds are open and there are no house rules to govern the Quantum String Theoriverse, one possible model for Creationism, out of a vast number of equally valid possible models, is an empty or ethereal quantum fizz for most of time followed, given the inherently biblical nature of that particular belief system, over the course of six days plus a day of rest, by exquisitely arranged matter for the last 6,000 years during which, for example, dinosaurs lived, died and were fossilised, and (with a bit of help from a visiting Newfoundland, and assorted worthy others I'd like to thank such as my literary agent, publishing company, and the publicity and supply chains, and all of their staff to name but a few) Pyror wrote a book about the house rules that allowed such a universe. And remember, crazy though this particular model may sound to many of you, with carbon dating appearing to validly show timescales of mega-millennia, but in this particular case (arguably) being something of a red herring caused by the pre-matter fizz, it is (arguably) one possible *valid* model.

Let's look a couple of other weird models (because everything is a BIT odd!): some good, some bad. Let's look at an apparently completely opposing view to Creationism (in which the bible is the literal word of God): atheism (in which, as you may already know, it isn't).

Please don't worry where this is heading yet!

Pyror, artisan firework manufacturer and, as will unfold later along with everything else touched on in this paragraph, symbol of canine optimism for the future, has lots of chums. He told me that while I was dead. What he didn't tell me was exactly how many of them if any are actively involved in the pyrotechnics, and how many are watching passively, perhaps with a beer and a hot dog. To what extent even Pyror himself is technically actually involved in an active way on a day-to-day basis will still be open to philosophical debate and will not preclude your own opinion, which is what Pyror wants, for reasons that will as promised also unfold later. One might even argue that it doesn't really matter in the slightest: the universe is what it is, and from our perspective a definitive right or wrong answer (about what goes on outside the universe), even if we could understand it, is about as likely as a ping pong ball in your game of Ping-Pongoverse leaping out of the box (and remember the box is the ping pong ball's universe), making you a nice cup of really hot tea or coffee, and having a chat with you about your interior decor. It's unlikely. Nevertheless we'll come up with a *great* answer to the preceding questions, in terms of what is within human and canine understanding, but there will be *some* limits.

A ping pong ball might look around its box and think to itself (and I defer to your own opinion on the intellectual capabilities of ping pong balls, rather than presenting this as fact), 'Aha! Matter started in Second 1. It is now Second 2. Before that the box exploded from nothing for some reason. I'm not sure why, but that's certainly what seems to have happened. The universe is therefore a second or two old. I'm a freaking genius. In fact, I'm obviously the most intelligent ping pong ball in the universe!'

I, Max Merrybear, submit that while indeed it might be the most intelligent being in its own particular universe, the ping pong ball has missed some reasonably important information, albeit I do not judge the ping pong ball harshly, for it really doesn't have any hope of ever working out *exactly* how its box came into existence and why it's there, as it does not have a dog to explain things in terms that even a ping pong ball could understand!

And it may just have to live with that.

Luckily, we don't have to. We have the option of reading *The Word of Dog* patiently, sequentially and conscientiously from beginning to end. And sometimes it can be good to philosophise or muse, as long as it doesn't stress you out too much, plus of course if you want to believe that your view of things is correct in every detail, no one has the right to tell you you're wrong.

So. Let's muse. What is an atheist?

Pleased don't worry where this is heading yet!

Quite simply, in Canine religion, it's someone who doesn't think Pyror or his chums are actively involved in the firework. That's slightly different from the traditional human atheist who generally confidently espouses the view, 'There is no God. The universe started in a bang. It must just have happened for no reason for some reason, and there wasn't anything there before, which makes perfect sense. Nothing outside of the universe is "alive" and if it was it couldn't "see" us.'

Canines, or at least the more intellectual of us, prefer to think outside the box (which is a deliberate Ping-Pongoverse-related pun in case you wondered), so we think of an atheist as someone who doesn't believe that any forces or life forms which may or may not live outside the universe have any influence on it, even if those canines are unable to offer a rational explanation for its existence. Most canines will accept that the universe does indeed look surprisingly like a firework, so the BONE of contention is really over whether anyone is watching it go off, and whether there's much they can do about it if they are.

At this point it's probably worth mentioning that there are very few respected (human) scientists working in related disciplines, perhaps even none at all, or at least none who I've yet met who were willing to talk about it openly to me, who believe that 'the universe' we live in is the only universe in existence, and that as a general rule of thumb universes are almost certainly teeming with life (unless they aren't and we really are completely alone), so by logical deduction the current scientific view is that life probably exists outside of 'our' universe. Whether it's particularly *intelligent* life, or whether there's life outside the greater continuum/omniverse, is a matter for reasoned debate and/or theology, or at least until everyone has read *The Word of Dog* by which time everything should be a whole lot clearer.

Pyrorists, ie those who believe and trust in Pyror, believe that just like you with the Ping-Pongoverse, Pyror can move back and forth through time, to way before the universe began (ie, in essence before he lit the blue touch paper) and after it ends (when the gunpowder and oofle dust, cf goofer dust, are all spent), and is willing and able to change the house rules so that after what to us might seem like an eternity, eventually the universe might become some sort of utopia, where war, disease, hunger, hate and tragedy fade to nothing. And as we demonstrated with the Ping-Pongoverse, if you think outside the box *this becomes a possibility.*

And there is an extra *very* important point here: Pyror lives outside time, as in outside the time that governs our universe, just as we essentially live outside the 'time' that governs the Ping-Pongoverse. The word 'eternity' therefore has no valid meaning outside the box, or outside the universe, as time ceases to be sequential. And I know that answer won't keep everyone happy, and least not yet, and especially not those of you with digital watches, but I submit that at some stage or another most of you have probably had the wondrous *feeling* of serenity and oneness you get from a timeless moment, whether up a mountain contemplating the vastness of nature, or perhaps gazing at a newborn face, so I offer this as further 'evidence' that timelessness is a valid concept (although of course the example does not attempt to propose how timelessness works outside of the universe: it's just an analogy) where 'eternal' loses the meaning it would otherwise have inside this universe. If you *haven't* had an experience of 'timelessness' yet, if at all possible you might want to get out and sit on a few more coastal hills where you can gaze at the ocean. Failing that you might gain similar enlightenment by lying on your back and kicking your legs around while someone rubs your tummy. Whatever suits.

Please don't worry where this is heading yet.
The Word of Dog **is NOT affiliated to any particular religion!**

Right. We've mentioned atheists. Let's do Christians. And Lord knows that's a woolly definition if ever there was one!

I'm going to start with 'middle of the road' Anglicans, mainly because I suspect that allegiance to one of the generic Abrahamic religions is where one or two of you have either come from or may head to at some stage, especially *if* a communal spiritual journey is what you *decide* to seek after reading this book, which will be entirely your choice unless you actually *want* me to tell you what to believe, in which case later on and throughout I'll tell you how I interpret things and let you decide if you want me to tell you to believe that, although it does beg the question, 'What is a Christian?'

Someone who believes in God? Someone who follows Christ? An atheist who thinks that there may be some merit in the socialist principles of tolerance and equality as apparently put forward by Jesus? An investment banker who has just collected a huge annual bonus tainted by the blood of third world farmers, but then tithes ten per cent of their salary back to Médecins Sans Frontières? A convicted thug who still doesn't accept he did anything wrong and who shouts at his wife and children when they don't conform unquestioningly to his view of things? A kindly old man lost because his wife lies buried in the churchyard, to where he now takes fresh flowers each week just like he did while she was alive? A young nurse who prefers to spend much of her spare time volunteering at a local animal sanctuary rather than partying the night away? A single mother who still wants her children to live a life of kindness and tolerance despite her own disappointments and traumas?

Several years ago, I went to a pet service in a modestly sized rural church, an ordinary middle of the road church, with nothing particularly unusual about the locale and with exactly the sort of people you'd see every day at work, in the park, or at the supermarket, and in that one service I met every single one of the above. And that was just a regular Sunday, albeit with a congregation swelled to forty-eight because pets were allowed that day. Including a Tamagotchi: one of those electronic pet things, and it too got taken for a blessing from a lady priest who looked and behaved, for all the world, just like the Vicar of Dibley. But I digress.

So. For the next few minutes, our Christian is Mrs Average Smith, married with two children at secondary school, one black (pedigree) Labrador, her middle manager husband usually attends church with her unless there's 'something going on at the golf club', although at least in this case it really does relate to golf rather than him having an affair with the 'free love aficionado' next door; and the family usually have one long haul holiday a year plus a couple of shorter breaks in the UK; in other words it's the sort of idealistically exactly average family that in reality one might expect to come across once every few years at the very most, if ever.

What does she, Mrs A Smith, believe? And what about Mrs B Smith the Baptist, or Mrs M Smith the Methodist, or Mrs H R C Smith the Holy Roman Catholic?

In fact, according to the Year 2000 version of the World Christian Encyclopaedia, and naturally even the counting method is open to heated debate and argument, there are 33,820 different denominations of Christianity, without even beginning to allow for the fact that within the denominations pretty much everyone maintains a subtle difference in the way they interpret things, or their images of heaven or how to live a 'good' life etc, just like (quite probably) every other religion, if only because the contents of people's heads are different so people tend to take subtly different inflections from the same information, even if received simultaneously from the same source. In fact the only reasonably safe assumption is that all the Christian denominations believe in God. Most of them believe he had a son called Jesus. That's right: *most*. Not *all* of them. Just *most*.

How on earth can everyone be right at the same time?

The 'easy' answer, which is not necessarily what I'm going to suggest *is* the answer unless you'd like it to be, is that maybe everyone's universe *seems* slightly different to the next person's universe because everyone's universe actually *is* slightly different to the next!

And because everyone's *universe* could in theory be slightly different, it follows that everyone could then interpret 'the heavens' differently as well.

But *how* might there be differences? Surely it sounds utterly absurd!

Pre-nerd-alert comforter.

Don't worry; a *very* simple explanation with a diagram follows the nerd alert, and it's got another ping pong ball in it, so you know it'll probably make sense ...

Nerd alert!

Different people's universes can be different because within a single unit of time all possible combinations of ping pong balls, atoms or energy strings can exist simultaneously (in theory and in the most extreme case, ie no 'house rules' and lots of 'boxes'), and that variation is possible because time is a dimension within probability, and matter is subservient to time, so within a probability dimension it is perfectly reasonable to have different universes existing concurrently, not just a couple of universes based on a few heads or tails here or there, but billions of trillions of squazillions of them. I'll paws for a minute ... Oops ... Annoying, isn't it ... BEAR!

Back to normal.

Anyway, that last nerd paragraph, for those of you who chose to peruse it, was a bit of a brain full, so let's put it in somewhat simpler terms.

Get another box (please). Not just any box: a box the same size as, yes, you guessed it, a Ping-Pongoverse box. Hurrah for the Ping-Pongoverse. All hail Pingpongor!

All we're going to do is place it next to Chum-1's box, ie the box in the middle of time, not that we have to choose the middle of time; you can choose a different second if you prefer: it is *your* universe. Without changing any of the ways the existing boxes are arranged, what I humbly beseech is for you, Pingpongor, to take an extra go on Chum-1's behalf using the new box but without changing your house rules. The only consideration is that for yours and Chum-2's boxes' layouts, Chum-1's extra box has to maintain the rules so, for example, balls still need to touch according to the rules which thou, Pingpongor, has made, although as long as the rules and the relationships to the other boxes are maintained, it's OK for the new Chum-1's box to be slightly different to the existing Chum-1's box, or indeed completely different if you decided you wanted a more anarchic Ping-Pongoverse with NO rules!

You might even decide to decree that there *has* to be some sort of difference between the two middle boxes. Otherwise everything might just be the same, which you might decide would do little more than waste space on your kitchen table, and would mean that you had to spend more of your time finding boxes and more of your hard-earned money on ping pong balls, but with the end result being something of a disappointment, whereas a subtle difference might be really exciting and entirely worthwhile.

Here's a pretty picture of what might happen. And please remember, on this occasion we aren't rewriting the past, so the influence is technically slightly different, although you could perhaps philosophise at your leisure about the exact nature of the influence arrows, so I'd strongly recommend not worrying about it for now.

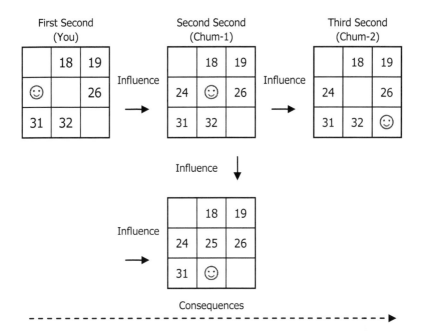

As you can see, the only difference between the two Ping-Pongoverses is that in the extra box the ping pong ball with the smiley face is in square 32, its concurrent 'equivalent' being in square 25.

But what *else* can you see?

That's right. For the middle time unit of the Ping-Pongoverse, there, before your very (mighty) eyes, are two very similar but nevertheless different Ping-Pongoverses occupying *exactly* the same unit of time.

Sound stupid?

Not convinced that it relates to the real world?

Want me to tell you something that some readers may find a little scary?

Sitting down?

OK.

Stupid or not, and quite frankly I have to admit that it does sound a little far-fetched, is that in essence that is exactly how quantum mechanics works, and for reasons I'll explain later in terms that I hope even Jazz can understand (by firing a lump of electricity and/or light at a thing with two holes in it), without quantum mechanics you wouldn't have electricity or light, both of which are used by plants in photosynthesis, so quite handy for the ecosystem and food chain.

No quantum mechanics equals no light, no electricity, no food chain and no life. Honest!

So. What should we do about it?

That's kind of up to you really: it's your universe. But it might just also be your heaven and hell, so my first recommendation would be *don't be an asshole*.

Naturally, that will come easier to some people (such as yourself) than others (such as the late Genghis Khan, but more of him later).

Let's have a look at a practical example. First of all we'll look at how easy it is for assoholism to arise in a single universe scenario. Then we'll explore the joy of concurrent parallel universes. It's 1944. Obergefreiter Wolfgang Porta, having just climbed out of his Panzer tank, hot engine now cooling after a busy day charging up and down the battle front, is talking to Rabbi Eli Moses.

'Hello,' says Wolfgang.

'Hello,' the Rabbi replies.

'I'm not Jewish.'

'Really? How odd. I am,' says the Rabbi, patting the yellow Star of David on his breast pocket.

'Oh. I'm sorry to hear that. Well, too bad; I'm afraid I don't want you in my universe.'

'Who says it's your universe? It's just as much mine as yours,' avers the Rabbi.

'No it blitzing well isn't. Right. That's it. Into this "shower" with you. Let's get you nice and ethnically clean,' says Wolfgang, waggling his Walther PPK at Eli.

'My life already,' the Rabbi says tritely. 'Asshole.'

Or it could go like this.
'Hello,' says Wolfgang.
'Hello,' the Rabbi replies.
'I'm not Jewish.'
'Really? How odd. I am,' says the Rabbi, patting the yellow Star of David on his breast pocket.

'Mind you, I suppose we do live in slightly different universes, albeit co-occupying the same time, so it's not impossible that while in my universe Jesus was the son of God, in yours he was just a prophet *sent* by God. Honestly, sometimes having parallel universes is a real gas!' says Wolfgang.

'Yes it is, and because you haven't murdered me, it's OK to joke about it!' the Rabbi replies.

'Ha ha ha,' they giggle in peaceful unison, embracing the diversity in their universes, swapping flowers, and sharing a non-abusive group hug with some well-fed smiling children from the Warsaw Ghetto.

Splendid. We now know everything we need to about both human and canine atheists, Christians and Jews. We don't know a *lot*, but we do know what we need to in terms of the contextual building blocks for the next level of knowledge we shall soon share. By all means go to Wikipedia if you want to know more about any of those religions. Or anything else for that matter. Wikipedia's great if you ask me.

But first, I think an interlude may be in order. Let's have a couple of very quick chapters on somewhat loosely related but nonetheless especially important matters. Then we can get down to the nitty gritty of what to do with our new-found information (from a Newfoundland ... it's a dog joke ... Jazz is laughing ... a bit ... sorry), and maybe we'll have a quick look at some other, previously incongruous or incompatible religions as well.

Or maybe not.

Who can say?

Obviously I could because as part of the editing process (and with a little help) I've had to (try to) learn every letter, number and punctuation mark, and check their positions relative to each other in the quest for literary perfection, but my publisher said I ought to have a cliffhanger at the end of the chapter to keep you turning the pages, and that the element of doubt over the degree of, in his words, the 'continued comparing and contrasting of conflicting creeds could be a commendable cliffhanging contrivance'. As if knowing the answer to everything you ever wanted to know isn't enough on its own to keep you reading.

Honestly. Publishers. Just what planet *are* they on? Maybe we'll find out ... And then find out how to get there ... Now *that's* worth reading about!

Chapter Five

Elizabeth

Mad-B came home sad today. He's a Quantum Proofreader by trade these days, where 'quantum' is in the legal sense rather than the scientific, although he's happy to let people think he's more clever than he is if they aren't aware of the somewhat more obscure alternative definition. The gist of what he does is to format, proofread and lightly edit reports produced by seasoned expert occupational therapists on the costed rehabilitation needs of people who've been injured or harmed in some way. You probably know the sort of thing: 'Tripped over a pebble? Grazed your knees and got the boo-boos? Want £500 compensation while the defence are obliged to pay our fees of £5,000 for two days work? Give us a call today!'

Only Mad-B works at the higher end of the scale on the messier claims. Claims generally from as little as £100,000 to around £10 million. And there's quite a bit of tragedy and social misery behind the scenes, and sometimes he cares more about the claimants than others, and sometimes there's the odd scumbag trying it on, but not often because usually they have quite messy injuries. And often, as with so many things, Mad-B's repeated exposure over time has lessened the effect it has on him. He still cares, but often it doesn't burden him for too long. But today he was sad.

The lady concerned, now 27 years old, had never known

her father and had a physically and mentally frail mother who lived in a care home in her native Poland. The young lady, we'll call her Elizabeth because that's an easy, anonymous translation, had two older sisters with whom she had lost contact many years earlier, although Elizabeth made the most of things and was happy living in England.

And about three years ago, she and her newlywed husband were involved in a road traffic accident. And when the occupational therapist flew out to Poland recently, where Elizabeth is now cared for, Elizabeth's only voluntary movement in any part of her body was that she could, on being asked in either Polish or English, wriggle three fingers on her right hand. She could not speak or even grunt, although she did, three years on, still grimace with pain whenever she was moved. And her only other response of any kind was to motionlessly shed tears whenever the nursing staff mentioned her husband, himself killed in the crash.

Every other day, Elizabeth's cousin comes to visit for an hour, but due to the shortage of staff and funds, that's often the only time Elizabeth gets to sit upright, and for the other forty-seven hours she's just a pair of eyes staring at the ceiling in pain and loneliness beyond that which anyone should have to endure.

At least the claim will mean that a proper care, therapy and pain management package can be implemented but, even so, forcing her to stay alive is one of the most inhumane things I can imagine, and I'm certainly comforted to know that Mad-B, who I know loves me more than I can express in mere words, would let me go to see Shadow with sorrow beyond measure but barely a moment's hesitation if I was in a similar crash.

And I don't think Pyror would want that life for Elizabeth either. I think he's probably taking notes, not only about Elizabeth but about all of the things that aren't quite right, and I think he'll be having a look at all of the house rules. And I think he'll pick a point in our universe's time, add a few new rules, and rerun our universe to see if it works better. And I think he'll keep trying until he thinks he's done all he can without removing the free will that would otherwise leave us all trapped in a gilded cage, which as mentioned in Chapter Four means he has to carefully moderate his *active* involvement within certain unknowable boundaries.

But what does that mean to us, here and now, in our imperfect universe, in our 'time'? Naturally we're going to discuss that more later in various ways and with varying degrees of subtlety or depth.

Firstly, however, I think it's time I proved the existence of Father Christmas; just in case some of you have lost faith since the days when you could see things with child-like clarity. Here comes the grown-up version ... and you might want to sit down: this is actually a serious discussion, not just some glib cretinous clichés that you might expect from one of the lesser mammals. You really are about to find out the truth behind Father Christmas.

Chapter Six

Father Christmas

Let's go back to Chum-1's Ping-Pongoverse for a moment. Pick a ping pong ball, any ping pong ball; unless of course your Ping-Pongoverse is an empty vacuum of nothingness, in which case perhaps you would indulge me briefly by playing the role of Pingpongor once more and commanding yourself and your chums to play again until between you you've deployed some 'matter', ideally a ping pong ball that survives for all three seconds, although it's fine if it moves around a bit.

Pluck, for now, your chosen ping pong ball from the second box and hold it mightily aloft, riven from its universe at your whim and command, although if you could avoid going 'Moohahaha' and bouncing it on the floor that would be great, as ideally it would help if you were still playing the part of a benign deity. That said, I suppose it works either way if you're feeling a bit mischievous or malevolent. Subject it to the sort of defilement you might expect of a drunken rugby-playing student if that's what helps get you through the day, although please wash and dry it carefully afterwards as necessary before continuing the experiment.

Now take a pencil and draw a faint zigzag line around it.

Now put it back exactly where it was in the Ping-Pongoverse, and preferably the same way up if at all possible, or it might get *really* confused!

Good god, ie you, Pingpongor the Good, do you see what you've done? Within the Ping-Pongoverse, apart from one ping pong ball now having the faintest of lines around it for some reason, nothing has happened at all. Apart from the line, everything is exactly the same throughout time and space as it always was, or so it appears to the ping pong balls inside that universe. They, that is to say Pingpongor's loyal subjects of the box, ie the ping pong balls, didn't even notice that one ping pong ball temporarily slipped out of the universe, because inside the box no one had time to see what was happening: that is to say at the start of one time unit the ping pong ball was there, and at the end of the time unit it was still there (although of course you, Pingpongor, know what happened because you live outside the box). And even though nothing happened, at least as far as the Ping-Pongoverse is concerned, at the end of one time unit only, equivalent to 5×10^{-44} seconds in the Quantum String Theoriverse, there was a faint line around a ping pong ball that disappeared unseen by the start of the next time unit. You, ie Pingpongor, actually did something that was ever so easy for you (as long as you are able-bodied or have a helper, or are just picturing the scene), yet inside the box it was too remarkable to see or barely even contemplate as legend, let alone understand.

In addition, if you like, you could have a house rule along the lines of: 'If I've drawn on a ping pong ball, I'll draw the same pattern on the corresponding ball in the boxes that follow, even if it changes position, until I decide that it feels like it's time for a change, so the pattern can last until the end of time if I, Pingpongor, think the ping pong ball is glad to be decorated.'

And that, despite much silly folklore to the contrary, is the essence of how Father Christmas operates.

He plucks the earth out of the universe for less than 5×10^{-44} of a second, during which he has all the 'time' in the world to visit by reindeer and sleigh all the good little boys and girls *whose parents aren't cynical unbelievers*. Obviously he has to do this several times, presumably at least once for each earthly time zone as it enters the dead of night, plus maybe a few million times to restock the sleigh or revisit any homes where some otherwise deserving recipient has decided to 'try to wait up to see Santa', but he could do it billions of times if he needed to and it would still be too quick to see.

And naturally, where someone *is* still awake, Santa *could* just deposit the gifts unseen, as he works in the space between two units of time, thereby also giving him the opportunity to use an exotic form of 'quantum tunnelling', as used in energy transference by plants in photosynthesis, to effectively teleport into and out of any heavily barricaded and chimneyless buildings, but he prefers to avoid confusing and disappointing those people who don't fully understand his modus operandi, ie by accidentally making them think 'that's funny, I'm sure I was awake but my presents just appeared out of thin air and I didn't see Santa; I'm a bit sad now, and also scared ... I think I'll scream loudly and inconsolably until dawn ...'

Why does he only visit the good little boys and girls (plus the occasional adult and other good sentient beings including dogs, dolphins and some of the lesser mammals) whose parents or other cohabitees aren't *cynical unbelievers*? Simple. Picture the scene on Christmas Morning, 5.30 am, and Little Johnny bursts excitedly into the parental boudoir where his parents were sleeping, albeit fitfully because of their filthy unbelieving shame: 'Mummy, Daddy, Santa's been and he's brought me a PS7 and ten games including Grand Theft Holoship Anthology III!'

'What? You little toe rag. We didn't buy you that, and Father Christmas isn't real, so you must have hacked into our online pension fund meta-database, stolen all our life savings and bought it yourself. Right. That's it. The Newfoundland puppy *we* bought you is going straight into a hessian sack and into the river!'

Santa doesn't want that. And you can probably imagine what I think about it. GRRR! So Santa only visits households or establishments where there aren't any naysayers who might cause trouble. Likewise, of course, Santa is really there to deliver the *spirit* of Christmas, not lavish gifts, so if you wake up happy on Christmas Day and there's something simple like an orange, some nuts and a lucky sixpence in your stocking, Father Christmas has probably been. If you wake up and there's a PS7 and ten games including Grand Theft Holoship Anthology III in your stocking that your parents or other carers swear blind they didn't buy you, even though there's a receipt from a toyshop in the parcel with their partially-asterisked debit card number on it, it *might* mean they are heretics who don't actually want a puppy, don't actually have a pension fund, and are hoping you feel so guilty for committing a theft that deep down you know full well you didn't actually commit, that you immediately offer to let them come and live with you when they get too old and incontinent to look after themselves, instead of being put into a low-budget rest home.

So, to summarise. Santa delivers Christmas spirit, simple

pleasures and love beyond time to all those whose parents, carers or significant others aren't *unbelievers*. It's obvious when you think about it. And like a faint zigzag line drawn around a ping pong ball, Christmas spirit needs a bit of nurturing if it's going to last for any meaningful length of time. And you've proved to yourself the possible existence of a life form such as Santa, an outside-the-box deity-like being, because you've been one yourself.

So relax. Santa loves you even if he can't always be there, so things will probably turn out fine in the end, even if waiting might sometimes seem like unmitigated suffering for what seems like forever. I think Santa will visit Elizabeth this year, even if all he can do for now is pluck her out of time for a moment and let her visit a parallel universe where she's happy and loved. I hope he visits you too.

Chapter Seven

FAQs

Rightio, after that interlude, I suppose the real question is, 'Now that we know a bit about ping pong balls and the rest of the universe, what are we going to do about it?'

One thing to note, if you would be so kind and before we get down to breaking that rather sweeping question into forty-two somewhat more manageable chunks, is that as we established earlier, assuming that you want to believe it, not that you have to as I'm just telling you the way things are in *my* universe, is that everyone's universe is (or is *potentially*) slightly different but occupies the same time using slightly different parameters (house rules).

You might therefore decide that as it's *your* universe you can do whatever you like, and if you're really nice and really lucky, maybe you can. I would, however, make one observation: the only difference between your universe and the next might be that in ten million billion years from now there might be a noise too quiet to hear in a microscopic forest too small to see on a planet the size of a fingernail orbiting a sun no larger than a cosmic fireplace in a dwarf galaxy too far away to notice. In fact, the difference between your universe and the next might be just *one energy string for one unit of time*, which in terms of what you can see would mean they *appear* identical. That, essentially, is the definition of a continuum: 'A continuous sequence in which adjacent elements are not perceptibly different from each other, but the extremes are quite distinct.'

So if you know someone who's an asshole, you can take solace in the distinct possibility that sooner or later someone else from another universe is going to poke them with a sharp stick.

That does raise an interesting point though: to our finite brains in our oh-so-short lives we may have to consider an almost unfathomably large number of universes before we manage to find one that's different in a way which we can actually see. I wouldn't worry about it too much though: realistically, it's impossible to even guesstimate roughly how many universes that might actually entail. Unless perhaps you're Pyror and live outside the box, so that at least you can see how many boxes there are. And while we're on the subject of short lives, I'd like to point out that based on a recent Kennel Club survey of UK-based Newfoundlands I'm 'now' (at the time of writing) 9.75 years into a 9.67 year life expectancy, which for some bizarre reason and despite relatively recent leaps in our longevity, still seems to be the lot of a UK-based pedigree Newfoundland, while less fortunate Danish Newfoundlands get only a lowly 8.0 years, yet conversely some miniature mongrels get twice that!

Mild nerd warning.

Dear nerds, yes, $1 <= n <= 2^{GOOGOLPLEX}$ would of course be technically correct in terms of the actual number of universes, but that's a *range* and does *not* count as a guesstimate!

Back to normal.

Right. FAQs.

1)   You said you died and then got better. Is that true?

Sort of, and if this were a work of fiction I'd just say 'yes' and then claim that my heart had stopped beating but Mad-B gave me the kiss of life then got it going again, or some trite nonsense like that, but Pyror said he wanted me to tell the truth, the whole truth and nothing but the truth about it so here, with a little bit of background information, is the truth.

Mad-B first met me when I was just three weeks and six days old and he was living in Bexley, South East London, trying to bring up his kids on his own full-time while working full-time and doing all the housework, when he had the first of three two-hour interviews from my canine mother Gemma's human life partners Margaret and Tony to make sure that Mad-B would be a suitable human to raise one of Gemma's babies. He passed the tests (obviously), so at nine weeks old I went to live with him.

Mad-B always says he thought that Margaret, now a life-long friend and the only person on his side he'd known less than fifteen years who got an invite when he married Mrs Mad-B, had doubts about where her baby was going, but that he honestly doesn't believe he'd have been able survive the many years of empty loneliness without his Little Maxi's eyes staring up at him until Mrs Mad-B saw me and thought, 'Ooh. He's nice. I wonder what his owner's like. Ooh. He's not too bad either. They both look like a bear. I wonder if they eat a lot and need a lot of grooming.'

Mad-B usually knows me almost as well as I know myself, and vice versa, and I know he loves me so much that if he really had to, even though he would cry rivers of tears for a year and a day, he would let me have pink liquid from the vet if I needed it.

Then, just over a year ago, when I was eight and a half, I was sick a few times one Friday evening and collapsed. My temperature was over 106 degrees Fahrenheit. I was all but unconscious, and thinking about food or even water made me wretch. I certainly couldn't stand up, nor support myself if two of my humans tried lifting me to my feet with blankets they'd clumsily managed to forcibly slide under me. I was incontinent of stool and urine and, having been stretchered to the vets on the Monday morning after an emergency home visit at the weekend, I spent three days on a drip before, now shaved to try to get my temperature down, coming home for a couple of days to say goodbye, with the pink liquid booked for the coming Friday afternoon, although deep down the vets thought it should have been much sooner and Mrs Mad-B was inclined to agree. Mad-B took the week off work unpaid from his new proofreading job, where fortuitously most of the senior management team were understanding dog lovers, and he spent hours just sitting next to me on the cold kitchen floor so that I knew I was loved.

And then, at around 6.00 am on the Friday morning, I groaned loudly in a way that meant upstairs Mad-B was plucked from his snatched disturbed slumbers, fearing my death throes. I then took a couple of steps and collapsed again before Mad-B could race downstairs to see what events were unfolding. I was still near to death but *at least I'd moved* so, with Mrs Mad-B herself now sporting a shoulder injury, Mad-B single-handedly lifted me back into the car, not easy considering the size of me, still a less than svelte 75kg after a week of not eating or drinking, and the fact he has a divarication of the recti (a sort of 'pre-hernia' condition) – ouch – and never at any other time in his life has he managed to lift more than 50kg! He then drove me to a specialist vets, Anderson Sturgess (now Anderson Moores) near Winchester, with the sort of private facilities that most humans can only dream of, a referral having been done by my normal vets, Endells of Salisbury, so that at least there was a chance the insurance would pay, albeit no guarantee.

The specialist vets suspected a brain tumour after the extended initial consultation (£300 plus VAT for the lengthy and invasive consultation alone, so about £200 per hour; and you may well have thought that normal vets knew how to charge!) but over the next four days, again with me mostly on a drip and (ahem) *cat*atonic, they ruled out everything in the book from Canine Flu to several different cancers to brain tumours to Lyme Disease to Meningitis and many more. The Computerised Axial Tomography (**CAT!**) scan alone cost £1,200 per hour. After four days the total bill for both vets was up to £5,600 (versus the £56.00 or so Mad-B had in life's savings), not to mention several gratuitous phone calls a day from Tony and Margaret, plus free advice from experts veterinary and otherwise they'd mobilised from the Newfoundland Life Partners Club, so with the insurance limit about to burst (and still no guarantee of payment were there any unseen technicalities in the small print) Mad-B decided I should come home again, the vets agreeing, even though I was still not eating, hardly blinking an eyelid and expected to die.

And then, to the utter absolute disbelief of the specialist vets, very unsteadily and with substantial help and patience from Mad-B, I staggered some 42 yards to the car: the jaws of some of the finest veterinary minds in the country were well and truly on the floor.

I was still very weak indeed, and still didn't eat for two more days, making nearly two weeks in total, and so what with my age and arthritis Endells still had the pink liquid at the ready, and I still needed a blanket slipped under me to help haul me to my feet and support me while staggering to the garden, but for the next couple of months with each passing day and with each passing hand-fed bucket of chicken, sausages, pâté, Wyke Farms Extra Mature Cheddar, pork pies and the like, I got a little bit stronger. And now I'm running around like a six year old, chasing balls which I never did before, and enjoying every day to the Max!

Did I die?

No. But only because Mad-B had faith in me against all the odds. My case completion notes (some time after discharge) from Anderson Sturgess signed off with, 'Max remains something of an enigma.' That's why I have faith in Pyror, even when it all seems a bit weird and sometimes mind-bogglingly far-fetched: I suspect even the vets wondered whether I'd been kept alive for a then unknown purpose by mysterious forces unseen.

And after two months of anxious waiting, with Direct Line (the insurance company) having first written to all seven vets I'd ever visited to look at every minor detail of my life history to confirm that the small print didn't preclude payment, a cheque turned up for almost the entire cost of the treatment, which as far as Mad-B's credit card provider was concerned was a massive relief. If memory serves, and it does, he drank a mug of ale and danced a little jig that evening, not unlike when the psychiatrist came to tea in Chapter Two; albeit in this case he attempted a rendition of Earth, Wind and Fire's 1979 smash hit *Boogie Wonderland*!

2) What happens when I die, as in actually finally die in times of my earthly life ending once and for all, rather than dying for a little while but then getting better?

Ah yes, 'Final death, what is it?' A question which mortal man and his dog have asked themselves for millennia.

The answer is quite simple really, or at least conceptually. Obviously it's going to be a *bit* of an unknown in the long term, ie once Pyror has revisited every interaction between every energy string for all the 10^{200} years we mentioned earlier, for however many concurrent universes he happens to be running, although equally obviously I'm sure many of you believe you know exactly what will happen, and I submit that it's good you're thinking outside the box, but in terms of cold hard logic all anyone can claim is that they know what they *believe* rather than averring that what they believe is definitely correct and precise in every conceivable respect.

What follows, however, and we'll add levels of detail to build our understanding as we answer more questions, until we have the most full, complete and definitive answer we could ever even dream of hoping to find, is what I understand happens to the more agreeable species of dogs. If you've been good, the same might happen to you.

Once Pyror is happy that he's written the best set of house rules for the *omniverse* that he can, and once he's run it as many times as he thinks he needs to make every being that ever lived, on every planet around every sun in every galaxy that ever formed, as happy as it ever possibly could be, which might well end up being a bit of a balancing act, he's going to make a 'home movie' of it and pop it into a hologramatic video viewer above his mantelpiece in 'repeat' mode, a bit like the hologramatic Ping-Pongoverse mentioned in Chapter Three, quite possibly as a romantic gift for Mrs Pyror with a little card saying 'I would move the very stars for thee.'

Furthermore the hologram, as we shall discover, will probably have different viewing options. What exactly that means, however, we shall have to cover later once we've learnt a few more things.

Remembering of course that in Pyror's lounge things exist outside of our universe, we then need to apply a degree of faith that in essence the hologram is how we can live 'forever' in perfect bliss, grateful for every 'day' of the life Pyror loosely defined for us even if he chose not to exercise absolute control over our deeds by creating an unbreakable rule for our every movement during every single split second of our lives.

This, essentially, as we shall unravel with increasing credibility and detail as we go, is the basic model for heaven which ultimately, by the end of this book, will enable us to satisfy both science and all religions, including and validating your own if you have one. That's why dogs don't go to war. It's a shame that humans still do in all of the universes I've heard about so far, so the human race must still be waiting for a few new house rules I suppose.

Let's start putting some meat on the bones.

3) So, am I going to live forever?

It's looking pretty much that way – as long as you want to. Conversely, however, it would seem entirely reasonable that there are plenty of parallel universes where at first glance all of the galaxies look identical, but in one galaxy, on one of the eight to ten planets (astronomers regularly like to fight over what constitutes a planet, dwarf planet, or satellite) orbiting one of the 100 billion or more stars in that one galaxy alone, say 'The Earth', there is one noticeable difference, ie whereas in the version of the universe in which you're happily reading this book, in the other universe you never actually existed!

Therefore you have an easy enough opportunity, in this universe, if you so choose, for you to beseech Pyror that in the final version of the omniverse, ie the mantelpiece hologram, or heaven as some people like to call it, he picks one or more universes where you're blissfully happy, rather than just one or more universes in which you never existed. I hope I don't need to mention this, but if you never existed, it will certainly diminish your chances of living happily forever!

There is, of course, a slight dilemma here, in that you might be feeling a bit down, so you might actually decide to beseech Pyror to *deliberately* fix it so that you never existed. Let's face it, most of us who aren't Newfoundlands have had days like that. Maybe even some Newfoundlands. I know of one who was found without water in the garden shed when a new family moved into a house in South Africa in baking hot weather. He's probably had bad days. Jazz certainly has.

It might be more productive however, although obviously it's entirely up to you, if you considered suggesting ways in which Pyror might like to think about fine tuning his house rules to improve your lot while gaining himself a few extra points in the 'best universe' competition, although of course you probably wouldn't want to go just barking orders if you think Pyror might feel a bit tetchy on the odd occasion, say if an unwelcome drunken uncle had popped in to use the toilet en route to a function of some sort and forgot to flush it, so it might be worth trying to offer some sort of thanks at the same time, by which I mean, 'Thank you for this lovely …' rather than offering Pyror the still-beating heart of a young virgin, which quite frankly seems a little dated in terms of techniques to appease or thank deities or deity-like beings, and I'm not convinced ever was that good an idea in the first place.

What I like to do is to keep a 'best and worst' list of the things I like most about life and the things I'd like to change if I could *and without being too unrealistic about it*. And I don't worry about things like feeling the need to have the same number of good and bad points because, with Pyror's help, I envisage some 'future' heavenly universe where my list *only* has good points. Here, in no particular order, is my current somewhat idiosyncratic list:

Good points:

- When Mad-B hugs me and calls me his Furry Superman or Special Little Gentleman.
- Ear and tummy rubs.
- Home-made chicken liver pâté on toast for breakfast.
- Sniffing things.

- Walkies where I get to sniff things.
- Sniffing things on walkies over hills near the seaside.

Things I'd like to change:

- Mad-B's intransigent attitude towards the effects of Theobromine on dogs (so I could get away with eating a bit more chocolate – I know that if I eat several whole big bars it might kill me but at my weight a couple of ounces might be nice once in a while, certainly wouldn't be toxic, and would probably just give me a buzz!).
- My depth perception (so I didn't have to concentrate so carefully on never ever accidentally scraping my teeth on anyone in my excitement when they feed their Furry Superman home-made chicken liver pâté on toast for breakfast and rub his ears and tummy before taking him to sniff things on a walk over hills near the seaside: I am convinced that if I was given a mouthful of pâté *every* day it would help to 'train' my depth perception, so I *could* achieve an improvement!).

Not on the list:

- My arthritis (because Mad-B buys me Metacam to stop it hurting, so there's not much else anyone can do about it at my age in this particular universe, so worrying won't help).

Sometimes I like to pause for a while and, as you may have realised, sit in the sunshine on a hill overlooking the ocean, one of my favourite spots being the top of the hill/cliff on the

South West Costal Path en route from Lulworth Cove to Durdle Door, secure in the belief that outside time Pyror will look at every energy string in every atom in every drop of water I can see. And as I pause a while to sniff the air, I think about my list, and I think about the universe as a whole if it only had good bits. But then I always try to flick my mind back to the universe I'm in at the moment, which for all its faults I'm rather glad to be in, and I enjoy feeling the sun here and now today. Maybe I'm just lucky. If you aren't already, I hope that soon you will be too.

4) What's that poem about a dog, you know, that famous one written in the dark ages when some humans still believed non-humans didn't have souls?

Aha. You mean Lord Byron's *Epitaph to a Dog* in which he juxtaposes his perceived unworthiness of egotistical humans relative to the holiness of a selfless loyal hound, written in 1808 in memory of his beloved Newfoundland, Boatswain, who had just died of rabies and who Lord Bryon reportedly nursed fearlessly just like Mad-B nursed me (albeit my condition was 'an enigma' rather than actual full-blown rabies!). It's interesting that you should ask, because purely by coincidence Mad-B managed to work the poem into his half-hour wedding speech. Yes, that's right, he publicly serenaded his new bride with a poem about a rabid dead dog. That, I promise, is the absolute truth. I think the poem is out of copyright, so here it is in full (and unfortunately, for some strange reason, some humans *still* believe only they have souls!):

When some proud Son of Man returns to Earth,
Unknown by Glory, but upheld by Birth,
The sculptor's art exhausts the pomp of woe,
And storied urns record who rests below.
When all is done, upon the Tomb is seen,
Not what he was, but what he should have been.
But the poor Dog, in life the firmest friend,
The first to welcome, foremost to defend,
Whose honest heart is still his Master's own,
Who labours, fights, lives, breathes for him alone,
Unhonoured falls, unnoticed all his worth,
Denied in heaven the Soul he held on earth –
While man, vain insect! hopes to be forgiven,
And claims himself a sole exclusive heaven.
Oh man! thou feeble tenant of an hour,
Debased by slavery, or corrupt by power –
Who knows thee well must quit thee with disgust,
Degraded mass of animated dust!
Thy love is lust, thy friendship all a cheat,
Thy tongue hypocrisy, thy words deceit!
By nature vile, ennoble but by name,
Each kindred brute might bid thee blush for shame.
Ye, who perchance behold this simple urn,
Pass on – it honours none you wish to mourn.
To mark a friend's remains these stones arise;
I never knew but one – and here he lies.

5) I've heard it said that the Large Hadron Collider, you know, that big particle accelerator thingy CERN built under the Swiss-French border, is going to explain the universe/omniverse once and for all and disprove the existence of one or more gods. Is that true? Please give your answer from both scientific and theological perspectives.

Blimey: that's a big FAQ! OK. Firstly, for those of you who aren't familiar with it, to quote a bit of their website at the time of writing, in this case reproduced with their kind permission which was given free of charge in what I inferred from their email was the first such request from a dog:

Warning: may contain nerdisms.

'CERN, the European Organization for Nuclear Research, is one of the world's largest and most respected centres for scientific research. Its business is fundamental physics, finding out what the Universe is made of and how it works. At CERN, the world's largest and most complex scientific instruments are used to study the basic constituents of matter – the fundamental particles. By studying what happens when these particles collide, physicists learn about the laws of Nature. The instruments used at CERN are particle accelerators and detectors. Accelerators boost beams of particles to high energies before they are made to collide with each other or with stationary targets. Detectors observe and record the results of these collisions. Founded in 1954, the CERN Laboratory sits astride the Franco–Swiss border near Geneva. It was one of Europe's first joint ventures and now has 20 Member States.

'The Large Hadron Collider is a gigantic scientific instrument near Geneva, where it spans the border between Switzerland and France about 100m underground. It is a particle accelerator used by physicists to study the smallest known particles – the fundamental building blocks of all things. It will revolutionise our understanding, from the minuscule world deep within atoms to the vastness of the Universe. Two beams of subatomic particles called 'hadrons' – either protons or lead ions – will travel in opposite directions inside the circular accelerator, gaining energy with every lap. Physicists will use the Large Hadron Collider to recreate the conditions just after the Big Bang, by colliding the two beams head-on at very high energy. Teams of physicists from around the world will analyse the particles created in the collisions using special detectors in a number of experiments dedicated to the Large Hadron Collider. There are many theories as to what will result from these collisions, but what's for sure is that a brave new world of physics will emerge from the new accelerator, as knowledge in particle physics goes on to describe the workings of the Universe. For decades, the Standard Model of particle physics has served physicists well as a means of understanding the fundamental laws of Nature, but it does not tell the whole story. Only experimental data using the higher energies reached by the Large Hadron Collider can push knowledge forward, challenging those who seek confirmation of established knowledge, and those who dare to dream beyond the paradigm.'

Back to normal.

I'll answer your question firstly, if I may, from a scientific perspective. The answer is this: people, and I am not talking here about the scientists involved, who believe that the Large

Hadron Collider will explain where the universe(s) came from and therefore possibly disprove the existence of one or more deities have, unfortunately, failed to understand its capabilities or have had it explained to them by such a person. One might infer from some public access 'Have Your Say' types of forum, such as on the BBC Website from time to time, that a small subset of those people get angry about it because they're sanctimonious egotists who think that anyone who doesn't believe *exactly* the same as them is retarded, even though they don't properly understand what they're trying to be so pompously self-indulgent about intellectually. Max Merrybear does *not* like sanctimony!

What essentially the Large Hadron Collider is *designed* to do, described here in terms of the Ping-Pongoverse as an analogy for the Quantum String Theoriverse, is to increase our understanding of what might have happened (and always remembering that you, Pingpongor, command time and can change the past) in some of the earlier boxes than we can 'see' at the moment. It might well also (ignoring the fact that allegedly it might rip a hole in spacetime and destroy the universe) allow us to speculate further about the house rules that seem to govern the way things are at the moment, and how much matter there is in the universe. But it cannot and never will be able to peer into what happened before time 'started' (ie to look at Pyror's empty kitchen table for example, or see the furrowed look on Mrs Pyror's brow when he told her that he was popping out to buy countless trillions of 'ping pong balls' and some gravity, and would explain everything later), and it cannot and never will be able to disprove the existence of any deities. This is because even if the Large

Hadron Collider could demonstrate what happened *before* time began, and it isn't *designed* to do that so that would have to be a matter of unexpected *luck*, that isn't a scientific proof for a negative situation, ie even if you prove the universe could easily have happened with no god (which I reiterate it isn't designed to), it doesn't *prove* that there isn't a god. Ultimately, however, *The Word of Dog* will allow any atheists reading this to carry on being atheists if they so choose. Likewise, everyone else will be able to carry on believing whatever else they may decide. Or they can choose something new to believe based on what they read here. And they'll be able to feel good about their choice whatever it is!

So, in summary, the Large Hadron Collider might well be useful, and there's a decent chance that it won't destroy the universe, and as I'm telling the whole truth I should confess that I'm actually mildly excited about what it might help us to discover and I've 'liked' them on Facebook, although that does raise an interesting point. Consider the effects on the Ping-Pongoverse if you, Pingpongor, had a house rule saying: 'If any ping pong ball builds a Large Hadron Collider and accidentally finds something I really don't want them to, like a Merrybear's Boson (a hypothetical particle which would be a 6.11-sigma demonstration of the probable existence of a deity, rather than a so-called god particle like the Higg's Boson), have a tantrum and throw all the remaining ping pong balls on the floor and stamp on the boxes.'

From this, logically we can deduce that in the Ping-Pongoverse if your ping pong balls displease you irredeemably you, Pingpongor, would be able to utterly destroy the universe in one unit of time, ie in one box all of your ping pong balls would be happily being guided by the house rules, and in the next they are blown to smithereens outside of time and space, ie on the kitchen floor. Therefore, in our universe, were Pyror to have a tantrum, say if the Large Hadron Collider caused a rip in spacetime, it would take him one of our time units, 5×10^{-44} of a second, to wipe out the universe, rather than messing about waiting another 13.7 billion years while the universe destroyed itself slowly by imploding at the speed of light. Awesome!

Next, as you requested, I shall answer from a theological viewpoint and therefore, if it's OK with you, for the sake of simplicity I shall refer to just a single god and in a masculine gender, although by all means take any such references in the sense of as many or as few gods of whatever gender(s), if any, you feel comfortable with.

The answer is either this: 'A God particle you say? Praise the Lord!'

Or possibly (and remember I like CERN – I'm examining different viewpoints here rather than expressing a personal opinion): 'Aha! So someone I've never met using a machine I've never seen and don't trust has got a picture that's meant to be something far beyond microscopic that's been splattered in a high speed collision and is supposed to make me suddenly believe that before the universe started there was nothing, and then it all just started happening for reasons they can't explain. Rather than messing around with a massive multi-billion Euro piece of equipment, why don't they just get a few boxes, some ping pong balls, read *The Word of Dog*, and learn to relax in the sunshine once in a while. Praise the Lord!'

Or even: 'Oh ... So there's no afterlife ... Right, I suppose

there's no point going to work on Monday ... I may as well end it all ... Hang on ... You might not have the full picture ... I want to LIVE ... Praise the Lord!'
I think that covers it.

6) I'm struggling to get my head around the idea that Physics is weird. I always thought it was perfectly normal. What's going on? In your answer, please explain and comment on the so-called double slit experiment, whatever that is.

Interesting. Originally I was going to answer, 'Um ... You thought Physics was perfectly normal? What kind of weirdo are you?!' But then I realised that it might sound like I was suggesting that you, my beloved reader, was weird, when in fact the weirdness originated from the person who originally asked the question. Funny things, words and time!

But who might have asked it?

All physicists, *especially* those connected with the Large Hadron Collider, know *and enjoy* the fact that physics is weird, and they probably all know full well what the double slit experiment is, so it doesn't sound like a physicist asked the question. Yet *normal* people don't think physics is *normal* either. Therefore, the question must have been asked by a nerd-curious person with some limited knowledge of the subject, perhaps just enough to scrape through Secondary School with a B grade.

Have you (the reader) worked it out yet? That's right: it was Mad-B! Let's answer, albeit with some **mild nerdisms in places** for the rest of this reply.

Put simply, the double slit experiment demonstrates beyond any logical doubt that little things like photons (ie 'bits of light') and even electrons (ie 'bits of electricity' – the little things that orbit the bigger things that live in the middle of an atom) can effectively *choose* whether to behave like waves or particles depending on the circumstances in which they find themselves. In the simplest form of the experiment, if you fire a laser at two close-together parallel slits in a thin panel, towards a screen behind it, rather than filing through in an orderly fashion, proceeding in a straight line as it had when fired, and hitting the screen in two separate places, the light forms a dispersed wave pattern where the 'waves' of light interfere with each other, a bit like having no ping pong ball in one square and a larger-than-life one next to it. When only one slit is open or, and this IS weird, if you do anything else to try to prove which of the two slits any given photon goes through, they behave like particles and there's no wave pattern, although the same amount of light gets through: it's not a case of half of it being blocked because one slit is closed – that doesn't happen! It's weird, but it's very much like they *only* behave like waves when they think *no one is looking*. And the *really* weird thing is that with the most sensitive (and expensive) equipment you can show that, if it thinks you aren't looking, one single photon or electron *on its own* can interfere with itself! A more flippant writer might have considered an onanistic reference to the photon's self abuse at this point, but I shall resist; this is a serious book ... Although I should probably point out here for the benefit of any readers below the age of 40, that in the old days 'self abuse' was a euphemism for the 'degrading sin' of pleasuring oneself sexually, viz onanism aka masturbation ...

Nevertheless, the next time a scientist corners you at a drinks party and tries to talk to you about the complementarity principle of particle-wave duality, you can reply, 'Ah yes; you mean when photons and electrons get caught masturbating.' Most scientists will leave you alone at that point.

Anyway, if you happen to be thinking, 'Piffle and balderdash, that's ridiculous; even if I am nerdishly strange enough to be able to fully understand why there's allegedly something weird going on, you can't seriously expect me to believe it,' alas I can only assure you that that's exactly what does happen, although quite frankly even I, Max Merrybear, found it hard to accept before Pyror explained it to me. The gist of the confusion revolves around Pyror's decision to have certain house rules that apply only to large objects, by which I mean atoms and anything larger, ie largely the stuff of classical physics, the stuff that the likes of Sir Isaac Newton and Robert Boyle usually dabbled with, but not usually and/or necessarily to small things such as electrons (which live at the classical boundary), photons and the other stuff of quantum physics. That said, if you get a bunch of somewhat meatier helium atoms *really* cold they can do some *very* weird things as well, like climbing up the sides of jars looking for a way out, and slowing light down so much that you could overtake it in a car!

Simply put, the rules in question might start: 'If and only if one thousand ping pong balls (energy strings) gather together at a density of one Merrybear, they shall behave thusly ...'

A 'Merrybear', clearly, is a new unit of density to describe the behaviour of energy within the dimensions internal to the visible universe. Cool ... I wonder if a Merrybear's Boson somehow facilitates the interface between the quantum and string boundaries, rather like the outside of a ping pong ball linking the 'stuff' inside to the table tennis bat. Now *there's* a thought!

But back to the double slit experiment, and don't forget of course that we're just using ping pong balls as an analogy with which to help visualise the stuff of very small pieces of matter. I'm not suggesting that two actual ping pong balls would form a wave pattern leaving one big ping pong ball and one empty space. And in the style of the literary world's finest tenuous links, I'm not suggesting that if a ping pong ball met an anti-ping-pong-ball there would be an explosion destroying both ping pong balls and releasing energy according to the formula $E=MC^2$, although ironically that is exactly what we would expect to happen (normally – there is some new evidence under consideration to suggest that it isn't *always* so, and quite frankly it's 'very exciting' so not surprisingly I'll come back to it later). Similarly, with fundamental 'particles' of matter, the spin of the particle is also of significant importance … Oh … just like with a ping pong ball if you put a lot of spin on it while playing Table Tennis!

The analogy is useful though, given a typical person or dog's tendency to look at the Eiffel Tower and think, 'Gosh that's big,' or to look at an ant and think, 'Gosh that's small.' Trying to say there are zillions of protons in a line of protons an inch long, and a squazillion energy strings in a proton is, to all intents and purposes, largely meaningless. A ping pong ball *means* something. Hurrah for the Ping-Pongoverse. All hail Pingpongor!

Of course back in Pyror's Quantum String Theoriverse we

can only really philosophise about what the rules actually are, given that I understand that a goodly proportion of life forms outside the box have superior linguistic skills to both dogs and humans combined, and most don't even use the word Merrybear in everyday conversation, unless perhaps talking about my conceptual visit, but probably not as a unit of density! I certainly aver that no theological conspiracy theorists should infer that the word Merrybear is regularly used in everyday conversation outside the box!

That said, we can at least continue to delve into how both small and large objects behave and, as particle physicists and the like have already begun to realise, in the end it's not going to be as simple as just having quantum and classical physics, neither of which do a *great* job of explaining, for example, the weirdness of the 'hydrogen bond' and no doubt a host of other vagaries at the transitional stage between small and large: Pyror could have as many 'grouping' rules as he likes, so two strings might behave differently to three and so on.

Incidentally, the hydrogen bond is yet another *really* weird but very useful thing, in that technically water ought to boil at 100 degrees centigrade *below* zero, rather than 100 degrees *above*, which of course would *seriously* mess up the chances of life forming: all the active life processes we currently know about stop long before it gets that cold, although by all means muse at your leisure on the concept of cryogenic life if you wish – anything's possible!

In essence, while you may well have been told at school that water is H_2O (ie made of two hydrogen atoms joined to one oxygen atom), in essence it isn't because hydrogen is a man-whore and forms a loose bond (affair) with a nearby oxygen atom from another water molecule while still having a strong bond (a marriage of sorts) to its wifely oxygen atom. These affairs come and go fairly regularly, but under ideal conditions water can behave almost like H_8O_4 in a sort of aqueous free-love hippy commune. Hydrogen *doesn't do this* with almost every other atom it bonds with, and certainly nowhere near as vigorously with anything else. Water is practically miraculous as it is without needing to be turned into wine!

Of course whether the fact that water is essential to life as we know it, and whether its rather promiscuous behaviour has in any way influenced certain species' own oft-alleged occasional impropriety, is a matter for philosophical and moral debate, but next time a hippy corners you at a drinks party and asks whether you fancy simulating a weak dative covalent bond for the evening, you can say, 'Eeew. No, Smelly Hippy. What do you take me for, an atomic tart? If that's how you feel I think you ought to go and tell that scientist over there to stop playing with his Large Hadron Collider and show you his double slit experiment!'

Where were we?

Ah yes. The weirdness of the universe.

Remember, it is entirely possible, although incongruous with many religions and by no means am I suggesting that you need to believe it, that Pyror entered the firework contest, aka the Best Quantum String Theoriverse competition, with no intention of life breaking out, but then noticed that it had, so maybe tried sending his only son to earth (and presumably perhaps also to any other planets with sentient life, possibly as concurrent manifestations) to explain that *as long as we don't mind trying to help a bit* it will all work out fine in the end, but maybe it all got misinterpreted because people weren't great at thinking outside the box, so tried to rationalise the message in a way that made more sense in just the regular three dimensions they were used to. Or maybe life *was* intentional. And maybe, although it's only as likely as anything else, the rationalisation was spot on and Pyror really does sit on a cloud with a long beard (in his kitchen, rather than our universe as such; remember, and I am being serious here, if you think outside the box it's perfectly reasonable to have clouds in a kitchen that can be used as seating even if we aren't entirely sure *how*). What is important is that you can only *speculate* about what happens outside the box. You can *believe* it and have *faith* in it as well, but *no one* can validly tell anyone else they're definitely wrong because that might make them an asshole and, as we established earlier, when Pyror is making his hologram, he's not likely to want assholes in it costing him valuable points, unless perhaps he scores extra points for every asshole who has a really crap or *hellish* eternity. Again, logically, we can only speculate and believe our interpretations on the subject of life outside the box, rather than discover and know … Except for anything presented as fact here in *The Word of Dog* … Those things are all true and may be zealously treated as undisputable factual knowledge.

7)   Jesus Christ!

OK. I'm sure you realise that technically that isn't a question but, yes, in the majority of belief systems (based on the total numbers of devotees), even atheism, there was a Jesus Christ, although whether he was the son of God, a prophet, or just some ancient philanthropist who thought it might be nice if there was a bit more social equality, is something that has, I understand, been the subject of a certain amount of debate and genocide.

There are very few people who believe fervently that no one called Jesus ever existed, and even fewer of those people are actually likely to read this book, but we do have to consider the possibility that in some games of Ping-Pongoverse there might be some boxes in which none of the ping pong balls were ever placed there as a message or maybe even a cunning jape by you, Pingpongor, to see what effect it had on the past and future versions of the Ping-Pongoverse, so equally there may well be some universes within the omniverse that never had a Jesus.

Who do I think he was? To be perfectly honest, it doesn't matter that much what I think, although naturally I will answer in minute detail throughout the book using both subtle and overt ways so that you can answer the question yourself in a way that leaves you comfortable and serene. The main reason Pyror asked me to put pen to paper (conceptually – I can just about hold a pen in my slightly webbed feet if I have to, but using a keyboard is *hard*, so we'll cover the area of the pen-to-paper logistics more later) was to help you decide for yourself what you'd like to believe: I'd be happier not taking any responsibility for your choices however, especially as my universe might well be slightly different to yours. You should only believe exactly the same as me if you really *want* to.

So, you're welcome to believe whatever you feel comfy with, but don't forget that so is everyone else (as long as they allow you the same courtesy in return and neither of you go around causing deliberate offence to each other or rearranging each other's ping pong balls without permission, and having first set mutually agreeable boundaries about how you might fiddle with each other's Ping-Pongoverses if you want to, which is fine as long as you *both* enjoy it).

Wordy sentence with several embedded observational or contrasting comments alert!

If you like to believe that Jesus was the son of God, that's fine, although of course his physical manifestation outside the box might well be different anyway (unless you believe otherwise), much in the same way as it being fairly unlikely (unless you're convinced that in this case the probability actually resolves to 'true', ie any required coin tosses give 'heads', which is of course entirely possible) that any of the friends or relatives of you, Pingpongor, actually resemble a ping pong ball (although Mad-B has just mentioned to me that he went to school with someone who through no real fault of his own was indeed the exact same shape as a ping pong ball, albeit a somewhat larger version, so anything's possible and something's certainly going on somewhere, likely or not, probably).

Back to normal.

Similarly, if you want to believe that Jesus was a prophet but not actually the son of God, that's also fine. Or you might want to believe that Jesus didn't have any particular prophetic abilities but you'd still like to heed the message that life would be more peachy if everyone stopped poking each other with sticks or throwing rocks at each other, which is also fine.

So, am I, Max Merrybear, a Christian? That really depends on your interpretation of the word, so it's hard for me to respond without knowing what you might infer from my answer. I don't like it when people or dogs poke each other with sticks and throw rocks at each other, so if that makes me a Christian that's what I am. If it means we have to try to kill each other because of a minor technicality of interpretation on a subject where an unabridged definitive answer inside the box isn't possible, then I'm probably not, so it's largely up to you if you feel the need to label me.

8) How many were going to St Ives?

Nerds and cats alert, and lots of them!

Firstly, let me briefly plagiarise an explanation of the Schrödinger's Cat experiment, as well as assuring you that I fully expect some people, such as those who decide not to read *The Word of Dog*, may decide to argue against the answer that follows. To those people, but without wishing to appear in any way ill-mannered, I say, 'Bite me!'

Schrödinger wrote (paraphrased very lightly for context): 'One can even set up quite ridiculous cases. A cat is penned up in a steel chamber, along with the following device (which must be secured against direct interference by the cat): in a Geiger counter, there is a tiny bit of radioactive substance, so small that perhaps in the course of an hour, one of the atoms decays, but also, with equal probability, perhaps none; if it happens, the counter tube discharges, and through a relay releases a hammer that shatters a small flask of hydrocyanic acid releasing cyanide to kill the cat. If one has left this entire system to itself for an hour, one would say that the cat still lives if meanwhile no atom has decayed. The psi-function of the entire system would express this by having in it the living and dead cat mixed in equal parts. It is typical of these cases that an indeterminacy originally restricted to the atomic domain becomes transformed into macroscopic indeterminacy, which can then be resolved by direct observation.'

The gist of this weirdness is an analogy for the quantum world, where a 'cat' can effectively be dead <u>and</u> alive at the same time, and not until we peep inside the box does the cat take the value 'dead' or 'alive'. To a normal well-adjusted person (including, as it happens, Schrödinger himself, who technically never actually averred that the cat would be both dead and alive at the same time despite the 'popular' assumption to the contrary), the cat would, of course, have a more unambiguous idea all along of whether it was dead or alive and when, if at all, its status changed (ie just after thinking, 'Ooh, that smells like Bakewell Tart ...' to which I am assured that the aroma of cyanide is remarkably similar) but in the quantum world, this weirdness is *what actually happens*, however ridiculous it sounds!

As if that isn't weird enough on its own there are, as we 'speak', multi-million dollar tests being planned to examine whether, in a similar way to the demonstrable Heisenberg uncertainty principle, where as mentioned if you look at something you change either its position or velocity, so you can never tell both exactly where something is *and* exactly how fast it is going (largely at a sub-atomic level – it is fairly unlikely that Werner Heisenberg would have had his speeding ticket quashed subsequent to the joke in Chapter Two), that if you 'observe' a radioactive atom it 'resets' the half-life 'clock' (which is at least possible in theory, however stupid it sounds, although my personal theory would be that technically the observational energy could restabilise the 'system' back to its 'default' setting, so it might just be possible) therefore ultimately, if you can observe the radioactive substance in the Schrödinger model fast enough and often enough, it will never decay and thus (within certain parameters relating to the actual longevity of cats and exactly what constitutes an 'observation') you can grant a Schrödinger's Cat relative immortality. This is *so* absurdly ridiculous, yet may actually be possible by virtue of changing the outcome by looking at something, that I'll wager if it happens scientists will turn to religion in droves for solace. I'd recommend thinking like a dog and turning to Pyror, in which case there's no particular need to wait for the outcome of the experiment to decide whether to link your beliefs to the behaviour of radioactive substances, as canine theology already has both possible outcomes covered anyway!

In my 'poem', however, the Schrödinger's Kittens existed in the Ping-Pongoverse, which according to the rules at the time was operating on a three coin toss, so based on 1 in 8 probabilities. However, based solely on my poem, and making no claim that the poem reflected actual events in any known Ping-Pongoverse, expanding or otherwise, and equating for now each life form to one ping pong ball, although accepting that it may be conceptually possible if somewhat unlikely that a ping pong ball could exhibit signs of life on generation of a suitable system of internal organs, and that they might then exhibit the characteristics of a new genus of Felix (Latin for happy or lucky) within the biological family Felidae and with a species of Felix Pingo-Pongus (which is therefore Latin for 'Happy Smelly Cat that goes Ping if it's Lucky' – which sounds rather like the name of a bad Chinese takeaway, boom boom), and so in the Ping-Pongoverse a Schrödinger probability may indeed actually, genuinely and unequivocally equate to one of our Schrödinger's Kittens ... Where was I? ... Ah yes ... The man was going to St Ives, his seven wives were also going to St Ives carrying a total of 49 sacks containing a total of 343 adult cats who were known to exist, possibly having survived a similar earlier experiment, although the history of the adult cats is of no particular relevance to our current calculation. I was also going to St Ives. Because in the Ping-Pongoverse things happen based on three coins being tossed, the number of kittens (which could be up to a maximum of 7 x 343, ie 7 per cat, which equals 2401) would therefore be between zero (a 1 in 0.875^{2401} or roughly 1 in 1.73×10^{139} chance) and 2401 (a 1 in 0.125^{2401} or roughly 1 in 2.61×10^{2167} chance). We can therefore represent the number going to St Ives, where there are 'me', one 'man', seven wives, 343 cats, and between 0 and 2401 kittens, according to the formula for the grand total being 352 <= x <= 2753 (because in our poem we were clear that we were enquiring about animate objects only, so the sacks should *not* be counted), or '352 is less than or equal to x which is less than or equal to 2753', where x has a median value of 652 (based on 1 in 8 of the 2401 kittens actually existing so, rounded to the nearest whole kitten, 300 kittens as the median value of kittens).

For those of you who hate maths, and there's a fair chance you'll have skipped this answer completely by now if you do, so you probably won't be reading this (!), 'median' is one type of average, essentially usually used when the most common occurrence is needed within a big set of numbers, often useful if you get a few big values that might otherwise skew the impression given by the result, so read 'average' if that makes life easier, albeit slightly more ambiguous.

Interestingly, of course, in the original poem the existence of the kittens was not open to variable probabilities, although as aforementioned due to some astonishingly sloppy English (perhaps deliberately and/or antagonistically so) it is actually unclear whether the man and his polygamous feline-orientated entourage were actually going the same way as the poet, and the whole thing therefore became utter nonsense. Logically, therefore, and this is genuinely not open to debate or opinion in this case, as we've just proved it to ourselves whatever universe we happen to live in, a cat-like Schrödinger's ping pong ball, Felix Pingo-Pongus, *actually* living in the Ping-Pongoverse still makes slightly more sense than the original nonsensical poem!

Back to normal.

I therefore recommend, although ultimately it's entirely your choice, if you are concerned that some of the things in this book still seem far-fetched, that you have faith in Pyror, given that as long as you have the cognitive abilities to process the information, you've already 'proved' a whole lot of weird stuff to yourself and, as we shall see later, having faith in Pyror doesn't mean that you have to be religious if you don't want to, even if it does mean you'd trust certain dogs a lot more than you'd trust many a human!

9) How can everything end up happy if we still eat other animals, man? Give your answer in terms of cows, lentils and trees. Peace, love and equal rights for bacteria!

Ouch! Please remember that I am a dog, not a 'man'!

Anyway, I suppose the easy but glib answer would be 'lots of different universes blah, blah, blah', although I have a funny feeling you might not be happy with that, plus you could argue that unless we're suggesting Pyror is going to have a ridiculously large number of different heavenly holograms (which he probably could if he wanted to, and Mrs Pyror was comfortable with the effects on their mantelpiece, and she didn't mind 'holo-dusting' becoming a more prevalent item on the list of household chores, hopefully shared equitably), that any afterlife you've managed not to get yourself erased from by being an asshole (which I can't imagine is something that might now happen to *you* but might be true of some people who *haven't* read The Word of Dog) may for now, by logical deduction and although subject to more detailed discussion and theorising later, be a compromise.

Of course there's always the possibility that in the same way you, Pingpongor, could have several versions of the Ping-Pongoverse on your kitchen table concurrently, it would seem entirely reasonable that Pyror might decide to have at least a few different versions of the Quantum String Theoriverse hologram on his mantelpiece at the same time, in a sort of grand heavenly indoor firework display and/or performance, although personally I *doubt*, and it is just a philosophical matter, that he'll want anywhere near $2^{\text{GOOGOLPLEX}}$ such 'ornaments' clogging up his home, especially as the vast majority of those universes would be terrifying, violent, boring, quasi-identical or all four and more. I mean, can you imagine Mrs Pyror's reaction when she popped into the lounge to tell Pyror that his supper was ready only to find him proudly displaying a holographic universe that amounted to nothing more than a violent 'splatter movie' (cf Large Hadron Collider) that went on for 'endless' millennia? Or what if Pyror's mother-in-law popped in for a surprise cup of tea, saw a seedily 'burlesque' hologram and fainted? Pyror would be on the naughty step for sure, so it probably isn't going to happen: the final number of retained Quantum String Theoriverses will probably be fairly limited compared to everything that's theoretically possible!

Nerd alert!

You could argue, and I'm happy if you do, that the most holograms that Pyror would need at *any given moment* to cater for all possibilities is $2^{GOOGOLPLEX}$ divided by 6×10^{250}, ie all the possibilities divided by the number of time units in the useful life of the universe. Please remember though, that because of the effect of the powers, this is still surprisingly close to $2^{GOOGOLPLEX}$ and would certainly still probably be impossible to write out in full. If any nerds feel like trying to work out a better scientific notation for the number, by all means do so: not that I'm at all sure that there is a more succinct alternative, but then mathematically I'm no Ed Witten. So, to summarise: the maximum number of possible universes is $2^{GOOGOLPLEX}$; the maximum number at any particular moment is $(2^{GOOGOLPLEX}) / (6 \times 10^{250})$.

Back to normal.

So, for now, let's just assume that Pyror may have a hologramatic firework display with 'several', or perhaps even 'many' Quantum String Theoriverses, but that he'll go for quality rather than outright quantity, and equally it's pointless stressing too much over whether the number he'll have will be one, one hundred, or one thousand million billion trillion quadrillion ... although not surprisingly we'll fine tune our answer later.

The key to the answer we seek, ideal though it may not be in terms of our own particular planet (except for the majority of dogs and humans as in *most* countries we're *both* near enough to the top of the food chain) is to have realistic expectations, which with a bit of luck in 'time' Pyror will perhaps manage to exceed spectacularly.

But, for example, not everyone is likely to win the lottery. And even if everyone did win a million pounds it would make little difference, because then (simplistically) once the money had been 'fed into the system' everyone's *current* house would probably cost roughly a million pounds more than it does now for the same thing, so no one would actually be able to afford, for example, a gardener or cleaner because they'd have to be extremely highly paid so that they could afford a mortgage or the rent. Plus of course not many people want to work as a cleaner or gardener once they've won the lottery, and if 'everyone' wins the lottery that does have to include those people currently employed in the generic domestic assistance and maintenance sectors. But that's a matter of theoretical economics (which in any case is at best invariably an acrimoniously disputed grey area) and therefore largely beyond the remit of this book, so with all due deference in this case you'll probably either have to trust my simplistic paradigm or muse on the consequences in detail for yourself.

Nor is everyone going to get to sleep with (insert the name of with whomsoever you would most like to sleep; hopefully your current partner if you have one, in which case you might like to add that to your list of things to be grateful for) and if everyone did sleep with (insert name) he or she might end up with a few psychological and medical issues, so he or she might well cease to be an aspirational nocturnal partner (or teddy bear), especially if you happen to be the three-billionth person to sleep with them!

So, if you're a cow, do your best to run around the fields in spring, and by all means pray to Pyror that in the final hologram you don't get violated annually to produce milk for humans, but likewise don't waste too much time wishing you were on a yacht cruising the Mediterranean because if you *do* ever make it onto a yacht in the Med, you may find that wishing yourself there was a mis-STEAK. Get it? Mistake ... Mis-steak ... I just thought I should point that out in case any cows are reading. I've known a few bright enough cows, but a lot of them have the brains of a yak. Not that a dairy cow would generally be used for steak instead of a more beefy bovine, but that bit was kind of hard to work into the 'joke', and I think I digress, plus I should probably confess that I have had more than my fair share of steaks in the past, so I'm not intending to sound judgemental apart from the odd guilt pang over my own eating habits ... Actually, with sincere apologies to any cows who *are* reading, I do rather fancy a Chateaubriand for tea now my juices are flowing ...

And if you're a lentil, expect to get eaten, but don't forget to try to take joy from the simple pleasures of doing whatever it is lentils do before harvest time. Lentilling, one assumes.

And if you're a tree, just be glad you can wave your leaves around, happily soaking up all that delicious carbon dioxide there is to eat.

And if you're a cow, lentil or tree, and you still hope there's more to life than being milked, or living in dread that you'll fall over and there will be no one around to hear you make a sound, take solace in the very real possibility that Pyror will decide to have at least one mantelpiece hologram free from flesh eaters, where cows are the top of the food chain and don't eat lentils or chop down trees to make grazing for beef cattle, and even at least one mantelpiece hologram that's just for plants where trees and lentils reign supreme. There may even be a separate one where Pyror keeps all the hippies, man, but I'll put some more meat on the bones of that a bit later as well!

Chapter Eight

MAQs

Hey, that's my name: MAQs ... Maqs ... Max ... Get it? BEAR. Oops, that's the second time I've said 'BEAR' (oops) for no apparent reason other than a spot of random hilarity.

So anyway, in this chapter I'll be proposing answers for your consideration to More Asked Questions, still without implying any specific significance to the order of their posing. We'll also cover a small handful of IAQs, or Infrequently Asked Questions; you know, the sort of things that only pedants and madmen might waste valuable time thinking about, albeit I shall endeavour to answer in such a way that they productively add valuable building blocks to our knowledge. Actually, come to think of it, I'm fairly sure Mad-B asked me all of the 'unusual' questions. Hopefully just to test me. Otherwise in all honesty I think I'd have to say he has 'issues'.

If you're still looking for explanations for anything once we've considered the answers to life, the universe and everything in all 42 formal questions or discussion points in this book, or if you just want to chat, by all means email me at maxmerrybear@gmail.com and I'll get back to you if and when I can. Or maybe catch up with me on Facebook, where I'm known as 'Max Merrybear', that being my actual name.

Incidentally, by emailing or otherwise communicating with me, you automatically invalidate any disclaimers to the contrary, resile from any related past or future injunctions or super injunctions worldwide, and assign copyright to/of/in all correspondence between us to me in perpetuity to use howsoever I wish including but not limited to publishing in or by any medium of my choice whether separately or bundled and bound with the offerings of other correspondents or contributors, with the entire contents edited enhanced or fabricated at my sole autocratic behest and discretion including but also not limited to your email address and/or any other personal or contact details supplied or otherwise derived whether accurate or not and whether or not anonymity has been requested.

Right. That's a nice easy sequel sorted out, although just so that it doesn't all seem too self-centred, on publication one of my Facebook friends will win a prize of up to **one million pounds!** Also, I promise not to use your communications for evil and/or world domination: Maxi is a *good* boy!

Let's get on with the chapter.

10) Further to the issues raised in Lord Byron's poem, please comment on the question of whether dogs and/or other animals have souls and what happens if they don't. In your answer please give reference to Rainbow Bridge, whatever, wherever and whenever that is.

I'll answer firstly, if I may, the second part of your question by plagiarising the anonymous work of poetic prose entitled *Rainbow Bridge*. However, if the author(s) would care to make themselves known, subject to reasonable proof I would be happy to send them some Good Boy Choc Drops, a bone, or, failing that, some cash.

Just this side of heaven is a place called Rainbow Bridge.

When an animal dies that has been especially close to someone here, that pet goes to Rainbow Bridge. There are meadows and hills for all of our special friends so they can run and play together. There is plenty of food, water and sunshine, and our friends are warm and comfortable.

All the animals who had been ill and old are restored to health and vigour; those who were hurt or maimed are made whole and strong again, just as we remember them in our dreams of days and times gone by. The animals are happy and content, except for one small thing; they each miss someone very special to them, who had to be left behind.

They all run and play together, but the day comes when one suddenly stops and looks into the distance. His bright eyes are intent; His eager body quivers. Suddenly he begins to run from the group, flying over the green grass, his legs carrying him faster and faster.

You have been spotted, and when you and your special friend finally meet, you cling together in joyous reunion, never to be parted again. The happy kisses rain upon your face; your hands again caress the beloved head, and you look once more into the trusting eyes of your pet, so long gone from your life but never absent from your heart.

Then you cross Rainbow Bridge together ...

Clearly that is a rather simplistic 'comforter', although no doubt eminently suitable for pre-school children. However, if quoting it to them, to avoid trust issues in later life I recommend that you explain to them that they can obtain the definitive answer when they're old enough to understand *The Word of Dog* fully, hopefully by say their early teens.

Rainbow Bridge does, regrettably, have two major failings. Firstly, it is *so* simplistically twee that it is unlikely to stand up to either careful scientific *or* theological scrutiny; and secondly, in the absence of a complete discussion, the possible inference might be that while dogs and other animals have to *wait* for a human to accompany them into heaven, humans who have pre-deceased their non-human life partners might go straight to heaven without waiting for their beloved pet. I dread to think what the writer's stance might have been regarding those animals who *haven't* 'been especially close to someone here'!

Dogs certainly go to exactly the same hologramatic heaven (or set of holograms) as humans, except perhaps humans or dogs who are assholes and who as mentioned beforehand risk getting erased from history such as those who, after finishing their detailed contemplation of *The Word of Dog*, still for some unfathomable reason insist on believing that pets are not allowed to join their human life partners in actual heaven if the arrangement be deemed mutually beneficial and desirable. As we shall see later, even pet rocks *can* go to heaven.

As Confucius may or may not have once said, the only thing that hasn't got a soul is an asshole!

Furthermore, the gist of what happens outside of time and our universe is that Pyror himself has a really large 'collection' of pets based on the forms of animals commonly found in our earth's history (and many more from other planets): at least two of every kind of animal in fact, including unicorns and magic dragons, and he would be in *floods* of tears if anything happened to them and he hopes he would *know a* (No-ah) person who could *hark* his pleas for help to save them if they were in danger.

I think there's something to that effect in many people's ethereal memories, so in essence deep down you probably already know the answer, although my understanding is that the message may have got confused by many over the course of millennia.

So, to summarise, if by the end of this book you still believe that good dogs don't go to heaven, I would strongly advise you to do some serious soul-searching to make sure that you feel comfortable and confident that you aren't likely to be erased from history. I'm quite happy to state that boldly and forthrightly, as there's a *fairly* good chance that people who *don't* like dogs won't be reading this anyway! I really don't like causing offence, even when it is the word of Pyror and your very soul (if you choose to have one) may be at stake, but just in case any dog-haters *are* reading this, I bring you glad tidings: it's probably not too late to repent and be saved!

11) Am I going to be reincarnated as the ghost of Lucius Vorenus?

I think, and I'm speaking as a Newfoundland here because that really is what I genuinely am, that we can all learn a lot about *human* nature from the fact that this question almost made it into the *frequently* asked questions section …! (Albeit not necessarily the most *important* questions of course, many of which come later once we know everything we need to in order to fully understand the answers.) In fact, as you no doubt may appreciate, there are actually three very major 'questions' merged into one here, ie one, are ghosts real, and two, is reincarnation real? For now I think we can treat your specific reference to Lucius Vorenus, a character deemed, according to the likely disclaimers of the BBC and HBO, to be a fictional character in their utterly magnificent triumph of a drama *Rome*, with any similarity between any other person whether living or dead probably being allegedly entirely coincidental if they did indeed have the usual disclaimers (despite having characters such as Julius Caesar), to be like a third major question which we could, I believe, if we so chose, put in the more general terms of: 'Can a character from my own or someone else's imagination, or someone I dreamt about of whatever gender whether or not it matches my current gender whether that gender be original or reassigned or hermaphrodite, become real?'

So, three questions in one: reincarnation, ghosts, and dreams and imagination becoming reality. What of it?

On first glance, many observers might think that whoever asked the question must be barking mad. But are they? (It was Mad-B, by the way: Mad-B asked the question. I rather think you may have suspected as much.)

One of the fundamental principles of having $2^{\text{GOOGOLPLEX}}$ possible universes (which is only actually the case if Pyror has a lot of 'boxes', ie universes, and decides not to have any house rules, but is *possible*) is that anything, and I really do mean absolutely anything at all, which is both utterly beautiful and utterly terrifying far beyond your favourite dreams and most hideous nightmares, is not only possible but (as long as it doesn't involve 'incorrect' time travel, which we'll discuss later, but isn't just a matter of simple paradoxes) in theory at least should actually be happening somewhere right this minute: if you can imagine it, whether asleep or awake, in $2^{GOOGOLPLEX}$ universes it is not only possible, but it is *bound* to happen!

At least that's the principle.

But how? It sounds utterly preposterous!

In terms of whether or not you decide that it might actually ever happen in your own personal universe, that's entirely up to your own beliefs and conscience (or those beliefs you decide to settle on after fully digesting *The Word of Dog*), unless of course those beliefs contradict Pyror's house rules in that universe whatever those rules may be if any, but that's a matter for you and Pyror to sort out between yourselves somewhen. Nevertheless, for those of you who would like to ponder the subject further, I shall propose a possible mechanism for the requisite transference of energy, memories and psyche, and it's actually quite easy, at least as a concept, assuming that Pyror hasn't got a house rule to prevent it in your particular universe should you have decided by now that the rules in your universe may be somewhat different to other people's universes, which Pyror could of course arrange if he so chose, so feel free to believe it if you wish.

Wordy sentence warning!

Firstly, I'm going to make things even easier for myself by saying that in this model, whether believable and valid or not, whilst reiterating that I'm not saying whether it's true even in *my* universe, as I claim no first-hand knowledge of such phenomena from my own experiences, is that all a ghost would be is a quasi-gestalt energy form of the original subject that hadn't itself achieved either full or partial reincarnation (or heaven, hell, quiet oblivion etc), where full reincarnation would solicit the greeting, 'Hi, I'm Lucius Vorenus,' whereas partial reincarnation would be more, 'Hi, in a past life I used to be Lucius Vorenus,' ie for partial reincarnation the host, Mr J Bloggs, is still the dominant psyche and just contains the essence and some memories that were once part of the make-up of an actual real live Lucius Vorenus from a parallel Quantum String Theoriverse where he wasn't just allegedly a television character in an allegedly fiction-based drama, despite the drama being based on the history of Rome as we understand it whatever the disclaimers may have said if there were any.

And another!

So, to summarise thus far, purely for the sake of this discussion, a ghost is a quasi-gestalt energy form, ie in this sense a collection of energies (to be discussed in a moment) that allows the essence of a being (whether a full collection of energy or only partial, hence 'quasi' in this usage as gestalt would normally be a full or greater such collection, but our collection will, as we shall see, have the potential to become full or greater eventually, synergistically with a bit of luck if it happened to actually occur) to continue to 'exist' in this universe for an unspecified amount of time, whether fleeting or enduring, after what appears to be the death of the physical body.

Back to comparative normality ... just about.

Essentially, again for the purpose of this discussion and as hinted at previously, being a posthumous quasi-gestalt energy form, or ghost, would then be a stepping stone to either oblivion, full hologramatic bliss, or perhaps carrying on for a bit in this Quantum String Theoriverse through reincarnation (not that we're saying this is what happens if you don't feel like believing it, which you might not because it might not be true anyway – we're just musing on the point and bouncing ideas around).

Now let's ponder (as in ponder, rather than sharpen sticks with which to poke each other) what *might* happen, while nevertheless carefully remembering that for every ten people you meet who either claim to be currently, or that they once were Lucius Vorenus, at least nine and a half of them are probably suffering from one or more serious mental health issues. Personally I'd be *slightly* more worried about the 1 in 20 who *might* be Lucius Vorenus, given that in Series I, Episode II, *How Titus Pullo Brought Down the Republic*, he calmly explained over an impromptu refined meal that he'd killed 309 fighting men but didn't bother counting civilians, then joked with his female host, a noblewoman, Atia of the Julii, Julius Caesar's niece, that the priests at the war temple gave him discounts on his offerings after the first hundred kills; now that's what I *call* a nutter!

So ... still just pondering ... Let's say that in a broadly parallel universe, just over two thousand years ago, that Lucius Vorenus, having actually lived in real life, died and his body was quickly cremated on a funeral pyre. But wait a minute ... What was that ... Just after he died ... Was it me or did, for argument's sake, 'some' energy strings leave each atom in his body ... Gosh, that'd be a thing ... I mean it'd be impossible to see unless perhaps there were a goodly proportion of them, and there would probably be no material change in his appearance ... But technically, and indeed *scientifically*, if those energy strings were able to stay together, maybe using a string or quantum mechanism as weird as the largely non-quantum hydrogen bond which we discussed in Question 6 and now know is essential to life itself, and quantum goings-on almost *always* seem more weird than classical, then doesn't that mean there's a mechanism for transferring what I shall, for want of a better phrase (a better phrase perhaps being 'a posthumous quasi-gestalt energy form'), refer to as someone's spirit?

Let's think about that further.

How do memories work? I don't know, at least not in any great detail. Do you? The standard biological explanation involves electricity, which we already know (based on the currently perceived house rules in all of the most popular universes) undeniably often works at a quantum rather than classical level (while exhibiting certain classical behaviours as well of course), fizzing around inside various nerves inside your head. However, and I'm just flinging this into the pot for you to consider, given the incredibly detailed and/or photographic memories some people have, on the basis of even one memory per nerve or cell it would seem pretty much impossible to contain all the information in a single brain (except in Mad-B's case of course – he'd have no trouble whatsoever fitting his memory into a single head, bless him). Actually, in many ways the fizzing quantum electricity would be a perfectly valid possible explanation as far as the ghost of Lucius Vorenus is concerned, although while we're here it may be advantageous to delve a little bit deeper to give you a more rich and varied suggestion of how the mechanism might conceptually work; that way you will have all the information you need to contemplate the matter fully with an open mind so that you can decide for yourself what to believe (and remember we're only discussing possibilities here; I'm not going to tell you what you should believe happens in your own personal universe – 'piffle and balderdash' is still a perfectly valid conclusion if that's what you decide to settle on!).

I think, given the whole thing of fizzing electricity being responsible for memories and for the sake of 'simplicity', that we can probably narrow down the sphere of our energy transference contemplation to what might happen inside the brain. That said, should you meet anyone who claims their liver (or big toe or any other organ or appendage for that matter) once belonged to the ghost of a fictional television character, I imagine that the transfer mechanism, if deemed by you to be reasonable, would be largely the same. The chances are I'd probably think that they were a fruit cake, but in theory anything's possible just as you Pingpongor have proven, and indeed there is allegedly some limited circumstantial evidence related to transplant surgery to support the possibility of other organs having memories, eg where someone has received a new liver and picks up memories or traits from the donor without even knowing who they were.

However, please note: it would make me a sad puppy if anyone ever used this as an excuse for infidelity, such as in the sentence: 'I'm terribly sorry I slept with my secretary; my penis must have been possessed by the ghost of a fictional television character.'

So. Maybe when someone dies 'some' (let's say one percent for argument's sake unless you'd prefer a different percentage) of the energy strings in their brain cells 'escape'. To an observer, nothing even remotely obvious would have happened: the brain (not that you'd normally be looking at it at the moment of death) would look the same because the cells would all still have the exact same three-dimensional form as before, and the weight would have decreased by maybe a half of one of Her Britannic Majesty's imperial ounces, ie one percent of the weight of a typical human brain (under normal earth's gravity, which for the purposes of this discussion I have assumed, although equally the *measureable* mass may be affected less or indeed not at all), so it'd weigh almost the same even if you did happen to know the before and after weight (and by the way, as an interesting but random fact, a guinea pig's brain weighs about four grams – less than one three-hundredth of a human brain – but nevertheless I personally have observed them showing a wide range of emotions and communicating using at least four distinct discernible phrases, 'let's mate' generally being particularly well enunciated), and certainly not enough of a difference that posthumous fluid transfer out of the general cranial area couldn't offer a 'rational' explanation for the weight loss, and at best the escaping energy strings, even if still bound together (as mentioned, perhaps by something conceptually similar to a hydrogen bond) would at the very least create nothing more than the most faint of temporary ethereal hazes given that they would (quite possibly) still pass through massive juicy atoms without any collisions, so even if you *could* see something you probably wouldn't consciously notice it, especially given that someone with their brain on display had just dropped dead in front of you.

And what about this: what if some of the energy strings were actually 'dark' in the sense that you couldn't see or *weigh* them, in much the same sense as the 'dark matter or dark energy' of which your average Physicist would like not less than ninety-five percent of the universe to be composed, given that the universe seems to 'weigh' far too little which, and we'll discuss it a bit more later, was one of the main drivers for building the Large Hadron Collider. It may well sound weird, absurd or impossible (although it isn't impossible, just weird and maybe a bit absurd) but *if* ninety-five percent of the universe is dark matter or dark energy, it's entirely possible that ninety-five percent of a Newfoundland (or person) is 'dark', ie matter or energy that you can't detect by any normal means such as sight, weight or gravitational attraction. And that, my dear reader, means that it is possible, however unlikely, that when you die ninety-five percent of your 'body' could escape unseen and lurk somewhere in this universe or some other parallel universe according to Pyror's house rules, until it rejoined some future collection of visible matter, possibly having interacted with some other energies in the meantime, thus possibly transferring the posthumous quasi-gestalt energy of Lucius Vorenus into a new host.

So, are you going to be reincarnated as the ghost of Lucius Vorenus?

Possibly!

12) I've heard that in some cultures Adam's first wife was called Lilith and that the Romans had a Jesus called Mithras. Discuss.

Oh me, oh my. It does sound like I should have explained the concept of concurrent alternative universes somewhat more clearly, although I suppose it would be foolish of me not to leap at the chance to underpin and expand that knowledge, and rather rude of me not to address your specific concerns, and crass of me in the extreme not to take advantage of the golden opportunity to pad my manuscript out a bit with some quality prose so that the cover price is more agreeable to my beloved readers. Allow me to paraphrase a little background information. WOOF! ... Sorry ... I just got excited and felt like saying that ... I suppose it makes a change from BEAR though ... Oops ...

Obviously these days, mainly since an episode of QI on the BBC several Christmases ago, a dog can barely settle down under a shady table in a pub garden without hearing some drunken philosopher or another expounding their own personal theories about Mithras. According to QI, Mithras and Jesus were both: born of virgins in a manger on 25 December (and of course we may wish to consider that in many universes 25 December is just the official birthday of Jesus rather than his actual one), died for man's sins but came back to life on the following Sunday (three days: Friday, Saturday, Sunday; Monday would be day four), had twelve disciples with whom they shared a last meal, and had devotees who symbolically consumed their flesh and blood.

A true scholar of Mithras would decry much of that as journalistic sensationalism, albeit they would probably then go on to confess that very little is actually *known* about Mithras other than that he was actually supposedly born from a rock (yes, rock), which might explain why some people started to question whether it was actually a *good* belief system, especially given that there was another religion just down the road which had a much more plausible immaculate conception ... And the other thing that's 'known' is that there are quite a few underground temples from which snippets of information have been surmised about ritual meals and the hierarchy of the organisation, plus that given the estimated number of followers it would have to be classed as a valid religion in much the same way that Jedi now generally is from a *popular* perspective. And in reality, as far as what's actually known about Mithras, that's pretty much it. Alas, despite the fact that there were just over 400,000 Jedi in the UK from the 2001 census, thus making Jedi more common in the UK than Sikhs, Jews or Buddhists, technically, despite the popular perception to the contrary, it still hasn't officially been recognised as a valid religion from an *official* perspective. This seems odd, as if I may quote from the website of the Jedi Church, 'The Jedi Church recognises that there is one all powerful force that binds all things in the universe together, and accepts all races and species from all over the universe as potential members of the religion.' That seems like a perfectly reasonable philosophy to me and, if conceptually God be the force referred to, entirely in agreement with many mainstream religions. But I digress ...

Similarly, many a dog will no doubt have grown tired of all the antagonistic philosophising over why the current bible is a little light on information about Lilith. If I may 'quote' the Dead Sea Scrolls (in translation): 'And I, the Instructor, proclaim His glorious splendour so as to frighten and to terrify all the spirits of the destroying angels, spirits of the bastards, demons, Lilith, howlers, and desert dwellers …'

But to summarise, in some Jewish folklore accounts (and remembering that Jesus was Jewish), Lilith was Adam's first wife, but because they were both made from clay she refused to be subservient to Adam, dumped him, slept with an Archangel who batted for both sides in terms of good and evil, and had lots of demon children. Still, almost all of us know *someone* like that! Oh, and allegedly she occasionally ate other people's children, although arguably that could just have been a convenient excuse used by people whose own children had had 'accidents' and who needed a quick alibi. It probably wouldn't stand up in court these days.

So, what have we learned? Probably the most important thing is that you, Pingpongor, could tamper with ping pong balls in different boxes within the same unit of time, so alternative Ping-Pongoverses can be concurrently correct, even if they disagree with each other. This is exactly what can happen, if Pyror wills it, in the Quantum String Theoriverse.

Wordy sentence alert!

Therefore technically, and according to Creationists, Adam could have had only one wife, Eve, or two wives, although the fact that Lilith is supposed to have agreed to one hundred of her demon children being allowed to die every day makes me think it sounds a *bit* unlikely, like an old version of a Creationist Quantum String Theoriverse where the house rules have since been changed and the 'game' rerun, so it's essentially the same as if it had never happened in the first place, although equally, given that we can imagine it, there may indeed be a concurrent universe where that's exactly what is happening, however painful and unlikely that may sound to us, whereas for Darwinists it's entirely reasonable that *at the same time* Adam was more of a conceptual person, invented by men of old to explain what at the time seemed inexplicable, and that the house rules (and ergo the Quantum String Theoriverse) are different still (or different because it's co-existing but in a completely different universe).

And another!

And of course technically Mithras *could* have been the first visit of Pyror's son, with Jesus being the second ... and Haile Selassie I of Ethiopia being the third incarnation if you're a Rasta rather than the second, although naturally many Rastafarians might be offended at the concept of Haile Selassie being the Third Advent rather than the second, although with acceptance of the concept of concurrent parallel universes I submit that Rastafarians may improve their chances of finding faith and inspiration within themselves, which is one of the cornerstones of their way of life; and just in case you happen to be from one of the more mainstream religions it's worth mentioning that the Rastafarian movement is indeed a valid monotheist Abrahamic Christian philosophy, cannabis being smoked according to their beliefs as a sacrament during bible study to heal the soul in what they see as the same way that some more mainstream religions use wine in communion, albeit some people do also misuse cannabis just as some people misuse wine. But I digress again ...

Back to normal.

Let's face it, Mithras never exactly caught on in a particularly sustainable way from most people's perspectives, so maybe in Mithras-compatible Quantum String Theoriverses Pyror decided to have a second go at sending his son to earth (and as mentioned earlier presumably lots of other planets where intelligent life likes to think it exists even if they haven't read *The Word of Dog*) so, somewhat ironically, in those universes Jesus was his own second coming.

The fact is that if Pyror willed it, just as if you, Pingpongor, willed a ping pong ball to appear out of thin air (by adding one to the box), or in Pyror's case willing a virgin to become pregnant, he could send his son as often as he felt like it, not just once or twice. Heck, the first time round Jesus might have been a Pachycephalosaurus. I doubt he'd have been a T-Rex because most of them were, by all accounts, more something to run away from rather than to follow devotedly, whereas conversely the average Pachycephalosaurus was by all accounts probably a fairly chilled-out herbivore, although quite possibly eating the odd insect on purpose as well, but you'd certainly have a fairly hard time trying to prove that *no* dinosaur could ever have been related to Pyror! Funnily enough, we'll touch on bits of this again later …

Finally, in a completely unrelated matter, I would very much love to hear about any successful marriages that evolve subsequent to the initial chat up line having been 'Do you come from a mainstream, Cretaceous, or Mithras-compatible Quantum String Theoriverse?' and the reply having been 'Aha! I note that you realise Tyrannosaurus Rex was a Cretaceous bipedal saurischian dinosaur, despite the popular misconception that it hailed from the Jurassic era. Let's mate.'

13) Dr Who can travel through time. Following on from your answer to Question 12, does that mean that Dr Who is Jesus?

No. Dr Who is more like Father Christmas. And on the face of it, this question initially sounded profoundly nonsensical.

Despite that, you have actually raised an interesting point, given that Dr Who is actually a Time Lord with two hearts and the ability to regenerate into a slightly different but still, at the time of writing, humanoid shape, which makes me think that he would be a formidable adversary in a Ping-Pongoverse competition, because he might be able to think of a house rule that allowed his ping pong balls to change shape in an interesting and useful yet orderly way, although my faith in you, Pingpongor, means that I firmly believe both of the Ping-Pongoverses would be held in the highest esteem come judgement day, so there's a very real chance of it going right down to the wire in a quantum finish decided by a Heisenberg uncertainty, ie we wouldn't know for sure who had won until the judge, Pyror himself for argument's sake, although perhaps a delegated doorman or pearly gatekeeper, looked at them, which does indeed fit with a lot of popular theories and lores on the judgement day of the afterlife, so actually you did indeed ask a remarkably valid, incisive and probing question: well done again!

Anyway, as he's a Time Lord, it follows that Dr Who is not technically human, and yet what with the whole regularly saving earth from menacing space varmints thing he, more than most, seems to deserve a place in the heavenly hologram collection, which demonstrates that non-humans can indeed deserve a place in heaven, and that it is therefore also perfectly reasonable for some of the other more favoured and agreeable members of the animal kingdom to go to heaven as well. Newfoundlands spring to mind. If you disagree, a Dalek might materialise in your wardrobe and exterminate you. Or, depending on which universe you live in, it might not of course, but personally I don't think it's worth taking the chance.

Hang on in there: the questions get more sensible quite soon and with hindsight you'll almost certainly be glad you had the patience, tenacity, courage, good looks and intelligence to plough through the weird ones!

14) Is music a solid?

I think we can safely say that, yes, it is. However, like light, it is *weird*.

I'm sure you may have heard (pun – music – heard – sorry) the classical explanation for how sound 'waves' travel, by making molecules bash into each other in whatever medium the sound is travelling through, and while it's not a bad analogy as far as classical physics goes, the explanation is, of course, a little simplistic.

Here's the new explanation, and in a way it's easier to understand, as unlike the classical model where deep down you know they're not telling you everything they should be, while at first and until you've had a chance to muse upon this answer it may sound even more odd, at least you can be confident and happy that this is what actually happens and that you aren't being lied to, because you aren't.

It's probably going to come as no great surprise to you when, in starting to answer this perfectly reasonable question, I begin by pondering the question, 'What is a wave?'

At the seaside, so we're led to believe, it's a wobble on the sea caused by a combination of wind and the gravitational effects of the sun and moon on the earth (and of course to a lesser extent the gravitational effects of everything else within the Oort cloud, and to a far lesser extent the gravitational effects of Albert de Quadruped on the planet Zog in the Magic Faraway Galaxy, not to mention everything else in the universe, albeit from our perspective that has no obvious discernible effect).

The effects on sea wave production of fish and marine mammals flapping their fins and tails is often, I feel, rather understated. And that, apart from the fact that science knows exactly *how* gravity manifests itself but hasn't got the faintest idea *why* it happens, hence another part of the perfectly understandable obsession with the Large Hadron Collider is, on the face of it and if you accept and factor in a few more variables such as the local topography of the seashore and seabed substantially influencing the significant variations in tidal ranges over relatively short distances, a perfectly adequate explanation for sea waves.

At the seaside, you can certainly make the also somewhat simplistic case for a wave being a solid, in that you can see it, splash it, and in Jazz's case ... Well, you already know what Jazz does, or at least you will if this is the second time you've read this book. There's a terrible secret of his buried deep within Chapter Nine for your amusement. But what about a *sound* wave? That's a whole different kettle of, er, fish ... I think that's called irony. Or a tenuous link ... Or at least it will be next time you read this, ie when you know Jazz's secret ... I think I'd make a *bad* novelist ... Unless perhaps tenuous links to future flashbacks are an acceptable literary tool, if indeed that's what this is …

How can I claim that a sound wave is solid? Simple really: in order to make sound, even someone so easily pleased that all they want to do is sit around all day designing Large Hadron Collider experiments, rather than doing something more useful but challenging such as integrating the behaviours of their Large Hadron Collider with the philosophy in *The Word of Dog* just like you are doing by reading this book, would agree that to make sound you need an exchange of energy. And what keeps cropping up whenever energy is mentioned? That's right: quantum energy strings, ie the stuff that when enough of it gathers together in a small enough space (conceptually – we'll explain it a bit more lucidly later for the nerds or anyone else who has concerns), it makes solid matter. OK, so for the sound to be solid in the traditional sense, ie being able to see it, touch it and drop it on your foot, it would probably need to be so LOUD that it would almost certainly KILL you, but just because it gets a bit more ethereal when it's quieter doesn't mean that it can't be a 'wavicle' in the same way that electrons undoubtedly are, at least in all the most popular known universes.

'But Maxi Bear,' I hear you say, 'Even allowing for a certain inferred levity in this particular response, everyone knows that sound can't travel through a vacuum. Therefore you must be mistaken. Therefore surely you must be mistaken about everything.'

I note your concern, and it's easy enough to get confused by these things, but in simple terms the gist is that for energy strings to manifest as sound, they need to make a quantum leap into the classical world, so essentially the same things can obey different house rules depending on their proximity to other energy strings, or in this case, matter. Sometimes the house rules may be slightly different, and sometimes they might be so different that it seems utterly absurd, as with the double slit experiment, which anyone can demonstrate to themselves on their kitchen table with the right equipment, part of which equipment is obviously a kitchen table, but I digress yet again.

Therefore, it is entirely reasonable that sound can be comprised of the building blocks of matter, yet due to the house rules, or quantum versus classical physics if you prefer, they may be unable to travel through a medium with no physical matter present, ie a vacuum.

Latin alert!

Quod erat demonstrandum, as nerds sometimes like to say when signalling the conclusion of a 'proof' ... Or do I mean 'Dog erat demonstrandum'? No – just joking – it would be 'Canis erat demonstrandum' in Latin, but I confess that it doesn't translate too well, viz 'Dog is to be demonstrated' ... I think I should refrain from attempting Latin jokes ... Sorry ... Although a popular rather than literal translation of QED is 'Which was what we wanted', so 'Canis erat demonstrandum' would become 'Dog was what we wanted', which has a certain ring to it ... If you have any further thoughts on the quality of my Latin jokes, by all means email me ...

Back to normal.

Anyway, after that, I think I'll go and see Mad-B about having a tummy rub!

15) Do you have much experience of hazardous materials? Only I just killed a bacterium by pouring bleach down the toilet, so now I'm worried that I might get erased from history.

Do I have much experience of hazardous materials? Oh my furry goodness me, yes. Unlike many canines I have significant knowledge of the subject.

Mad-B, among his many other talents, has an NVQ Level 4 (about the same as a university diploma or perhaps foundation degree), which was largely plagiarised and fabricated for him by his employers without even referring to Mad-B for part of the 'evidence' they fabricated, in 'Certificate of Technical Competence: Managing Hazardous Waste Transfer', as for a couple of years he looked after a hazardous waste transfer station, on the same site but separate to a civic waste facility and landfill site. The hazardous waste transfer station was entirely self-contained within a fenced off area some sixty yards by thirty, with a cornucopia of colourful signs letting people know just how exceptionally unwelcome they all were, and with tall hedges to screen it from not only public view but, conveniently, the security cameras. That made it much easier and cheaper, for example, to deal with organic (ie carbon-based) peroxides, many of which are explosive, by carefully tipping them onto the rough ground behind the warehouse and urinating on them, rather than spending a small fortune handling them and dealing with them by more conventional and safer means. Of course Mad-B was always *slightly* worried that his penis might explode, but that never actually happened.

An additional benefit of being shielded from the security

cameras was that Mad-B could take me to work occasionally, where I was able to study the basics of practical chemistry, such as the fact that manufacturers of household bleach add a small amount of sodium hydroxide to the sodium hypochlorite of the bleach so that if a small amount of something acidic is added to the bleach, such as by someone in their toilet, much of it will generally try to react with the sodium hydroxide and therefore avoid a lot of the potentially toxic release of chlorine that might otherwise occur. What *was* funny though, and I confess that I do feel contrite for finding it amusing, was if Mad-B was bulking up commercial and industrial bleach, and some wag had put battery acid in an empty container, and a *large* amount of acid got added to a 200 litre drum of concentrated bleach; there'd be vast plums of deadly chlorine gas billowing out of the doorless warehouse for ages and blinding the forklift driver. Talk about slapstick comedy – sometimes I rolled on the floor barking so much I thought I was going to actually laugh like a human! Especially if the forklift driver happened to be trying to store a few drums of volatile CFCs (the stuff that destroys ozone) on a pallet on the top layer of racking, with their lids off to let it evaporate at no financial cost, skidded on spilt oil in his panic and crashed into another pallet of concentrated nitric acid. There was nothing quantum about the fizz when that happened!

Equally funny would be when they were bulking up 2½ litre glass bottles of chloroform into a 205 litre drum and the helpful hands-on boss from head office would 'accidentally' drop one on the warehouse floor, then apologetically ask a new member of staff, such as Mad-B soon after he started, if they'd mind getting a dustpan and brush. On bending down to sweep up the debris, the minion would inhale the dense fumes largely collected at knee level and spend the next half an hour stoned, while their colleagues pointed and laughed at them. Hilarious!

Of course all these japes and misdemeanours had to take place after 3.30 pm: every so often the Environment Agency did a surprise inspection, and they became quite tetchy if, for example, they turned up and a driver was smacking holes in aerosols with an ice pick for 'shits and giggles' (his words, not mine, although it is true that he both giggled and looked slightly nervous while doing it), but they never arrived after 3.30 pm as they needed time to get home for their tea, plus they knew that the site might be shut any time after 4.00 pm. Funnily enough, if there was no eco-crime on the to-do list, the site generally closed at 4.00 pm on the dot!

Still, to answer the second part of your question, are you going to be erased from history? For killing the germ, no, you aren't going to be erased from history. If Pyror wills it so, he could always have a hologramatic heaven where germs are the largest organisms and reign supreme, content with their happy lives. Might you, ie the person who asked the question as opposed to the beloved reader of mine now contemplating the answer, get erased from history for asking such silly questions when you could be trying to make the most of life instead? Let's hope not. And of course I do defend your Pyror-given right to be stupid if you so choose, and I applaud your bravery in leaving no stone unturned in your quest for spiritual enlightenment. Well done!

Hang on in there: the questions get more sensible quite soon and with hindsight you'll almost certainly be glad you had the patience, tenacity, courage, good looks and intelligence to plough through the weird ones!

16) I've built myself a flying saucer out of wattle and daub, yet I am unable to fly it as I have no engine. Can you help?

Yes indeed, although given that you appear to be suffering from some form of cognitive impairment (and again, as always, any references to impaired thinking refer purely to the person who asked the question, usually Mad-B, who I nevertheless love despite his foibles, rather than the readers of my work who I assume are in no way cognitively impaired unless they themselves aver otherwise) I should probably point out that a wattle and daub flying saucer is unlikely to be either sufficiently airtight or robust enough to withstand the rigours of interstellar or indeed intergalactic flight, so I'd strongly suggest that you just use it to get to The Moon and back, The Moon in this case being the colloquial name for *The Moon Under Water* public house situated in Leicester Square, London, and some other cities for non-UK readers, the eponymous venue being an 'ideal' pub as described by George Orwell in a 1946 essay, rather than our relatively nearby celestial body of the same name.

As I'm sure you may or may not realise, flying saucers work in a very similar way to helicopters or aeroplanes, except that rather than using air as an integral part of the lifting mechanism, they use gravity. In essence, a helicopter's blades or an aeroplane's wings are shaped so that the air has further to travel over the top of the wings than it does under the bottom. This causes a much lower pressure on top and lifts the craft almost as if by magic and/or witchcraft and/or, as you might say, physics. A flying saucer works in much the same way, only instead of passing air over various bits of the craft, it activates an anti-gravity force field around the craft with a pinpoint 'hole' in a carefully selected and controlled position so as to allow gravitational attraction 'in' from only the intended destination. The craft is therefore 'sucked' towards its destination.

Obviously, if you're aiming at a far off planetary system or galaxy, the attraction will be very small to begin with, so the acceleration would be worse than a French eco-warrior in a 1960s Citroen 2CV, so to overcome this problem the best flying saucers harness the gravity not only from the object in this universe but also a number of quasi-identical (for the purposes of the flight) parallel universes. The trick with getting good performance from flying saucers is therefore to harness the gravity from as many parallel universes as possible, while remembering of course that gravity is caused by a house rule rather than it necessarily being something tangible that you'll be able to explain with a Large Hadron Collider, which is why until they read *The Word of Dog* scientists are likely to struggle to find a rational explanation, unless of course the house rule does indeed also have a discernible physical manifestation. As I'm sure you will have no doubt noticed, the practical need to access quasi-identical parallel universes does complicate the design of the flying saucer's drive 'engine' significantly.

Nevertheless, if you're operating your flying saucer based on a simple single-universe drive, you can always use a nearby star (such as, in our case, The Sun) to give yourself a bit of a turbo-boost to start, and it's usually perfectly adequate for normal interplanetary commuting within your own solar system. For obvious reasons however, ie the temperature, don't forget to calculate a hyperbolic trajectory around the nearby star, rather than heading straight into it! Likewise, if travelling interstellarly, you'll find braking by harnessing the gravity of an object now behind you a lot easier with a pan-omniversal drive, so if you're flying in a cheaper model you need to be fairly careful not to smash into your destination planet at near to the speed of light, which never augers well for space cowboys in their wattle and daub flying saucers.

17) I'm afraid I'm not entirely happy with your answer to the previous question. You told me what to look for in a flying saucer's engine, but not exactly how to build one. Why should I believe anything else you say?

Um, dude, you're the one who just asked a dog how to make an engine for a wattle and daub flying saucer ... However I'll be happy to answer your current question in two parts.

Firstly, the information imparted to me by Pyror related mainly to the answers to how science and different religions can stop being unpleasant to each other and learn to live in perfect harmony (where problems have been seen to exist, and not implying that you personally or any belief system to which you subscribe have anything to do with any such problems whether real or imagined). Explaining the general directions in which you may need to undertake further research with reference to the flying saucer's 'engine' was just a theory of mine which I was happy to share with you, although I submit that it is the only rational explanation. I don't actually know *exactly* how they work in terms of what parts to ask for from your local hardware store. Remember, I am a DOG! That said, by the time you've read this the CIA will probably have started work on the practical applications of my theory, so you could always send them an email and ask for a copy of their research. Mark it: 'Area 51 Anti-USA Bomb Plot Terrorism Jihad Secrets Exposed' and I'm sure they'll get back to you. It might be when you're least expecting it though. And they'll probably want to show you around Guantanamo Bay rather than Area 51 … Maybe just mark it 'Help! UFOs!'

Secondly, give it a month or so from publication of this book and I'll wager that there will be unauthorised 'The Word of Dog Anti-Gravity Drives' appearing on eBay, although if I were you I'd wait until I can get my authorised merchandising guys on the case (including but not limited to male, female, hermaphrodite, transgender and trans-species 'guys').

I would add one final disclaimer to both this and the previous questions' answers of course, and that is if it transpires that gravity is caused solely by ethereal house rules rather than anything tangibly physical, then the only way anyone will successfully be able to build an anti-gravity drive will probably be by beseeching Pyror (howsoever he manifests himself unto you, his demi-god, Pingpongor). In other words you may as well put an empty cardboard box in the space marked 'Anti-Gravity Drive' in your wattle and daub flying saucer, get down on your knees and pray like the Pope on Easter Sunday!

Hang on in there: the questions get more sensible quite soon and with hindsight you'll almost certainly be glad you had the patience, tenacity, courage, good looks and intelligence to plough through the weird ones!

18)   What effect if any does blasphemy have?

The first and most obvious effect of blasphemy is that the engine in your wattle and daub flying saucer may suddenly stop working. Depending on the position and trajectory of your flying saucer at the point at which any blasphemy-related admonishment was to be meted upon you, this may result in one of three consequences.

One, disappointment. This is most likely if your flying saucer is parked in the garage and you aren't in it: the consequence being that next time you want to use it, you'll find that it doesn't work.

Two, overheating. This is most likely if you were just popping over to Mars for the day, perhaps to blast a few of the space locusts from the terrifying 1967 film *Quatermass and the Pit*, starring Andrew Keir as Professor Bernard Quatermass. Upon your anti-gravity drive failing, sooner or later you'd probably get sucked inexorably into the sun (or some other sun billions or trillions of years later if your speed and trajectory were sufficient to escape the Oort cloud). As mentioned earlier, the sun is quite hot, or at least it is in my universe. And you'd probably find the price of water *astronomical*. Boom boom.

Three, stress. This is most likely if you had one of the better anti-gravity drives and were popping over to the planet

Zog for lunch with Albert de Quadruped, albeit unless you can afford a tachyon booster kit and therefore travel faster than light it would still take quite a while: several years in fact, and that's if you could accelerate reasonably quickly to the speed of light. If you manage to accelerate at 10g (for example the equivalent of doing 0 to 500 miles per hour, or 800 kilometres per hour, in 2¼ seconds), which is a little bit more than a fighter pilot can generally withstand on a *very* temporary basis, it would take about five weeks to reach the speed of light, so realistically you're probably going to want to spend a year or so accelerating so that it's nice and comfortable, after which you could cruise effortlessly at the speed of light for a few years, then spend another year decelerating. Except that when you turned the anti-gravity drive back on and engaged reverse, you might find that it didn't work.

You would then face a fairly stressful six months before you smashed into Zog at the speed of light (unless your auto-navigator failed to compensate for the variation in Zog's orbital position based on your amended velocity, in which case you might well miss it altogether and just die of old age in the vast cold loneliness of outer space): it's roughly six months but quite frankly I've got better things to do with my time than indulge in the differential calculus needed to work out an exact figure, especially given my age and the earlier data on Newfoundland longevity, although I'm happy to print the answer to the differential calculus problem in the sequel if any nerds feel like emailing the calculations and answer to me but, as you may have realised, I digress ...

As mentioned previously, crashing at the speed of light (almost 671 million miles per hour) generally augers poorly for both wattle and daub flying saucers and the space cowboys who pilot them: survival is certainly generally more complicated than the 'tuck and roll' method of breaking one's fall recommended by a space hobo in the otherwise superb series *Futurama,* in the episode *30% Iron Chef*, the series numbering being somewhat confused between different continents and therefore not quoted here, nor indeed relevant.

A final consequence of blasphemy, unrelated to matters of personal transport, is that while Pyror usually likes a convivial laugh and a joke as much as anyone else on either side of the box, if you upset him too much he might conceivably decide to erase you from history and replace you with someone more worthy, and yes, with the possible exception of the somewhat simple scientific and theological conclusions that were the primary motivation of the visiting simulant (read 'cyborg' if that's easier, or 'robot' if you're still baffled) I do mean exactly like in *Red Dwarf*, Series Five, Episode Two, *The Inquisitor*.

19) It makes Mr Ibbs angry when you talk about tachyons. Tachyons are theoretical subatomic particles that are supposed to travel faster than the speed of light (and no slower). The speed of light is the fastest speed there is. You're a moron. Everyone's a moron.

OK, technically that isn't a question, at least not in the grammatical sense, but I'll count it as a valid discussion point and answer it anyway; not that I know exactly who Mr Ibbs is, but if the rumours are true he's a gentleman of sorts who lives in Mad-B's head in the year 1928 and tells him to make dead people, and all will be revealed in due course in Mad-B's work of fiction *A Pat on the Head* as mentioned in the section entitled 'Also by Max Merrybear' in this book's preamble … although I digress once more … just how *do* some writer's stay on topic?!

So, tachyons and the speed of light or, more to the point, going faster than it. In fact, let's take the rest of this paragraph for a quick overview of tachyons, so if you know all about them already feel free to skip to the next paragraph where it says, '**Until now.**' Right. Here we go. Tachyons are useful for science fiction writers to justify anything to do with time travel or going faster than light, but which no one can confidently argue against because no one really understands them fully, or at least no one sane. And in the 'real' world, that's been their only use ever.

Until now.

Unto you, this I put: could not the mighty Pingpongor, if he so chose, allow a solitary ping pong ball, perhaps coloured a slightly different shade of almost white to the rest of the ping pong balls, to move two or more squares per 'second', but never *less* than two so it always *had* to move, while restricting coupled or pure white ping pong balls to a maximum of one square at a time? In effect, conceptually, what you, Pingpongor, would be doing is setting a maximum speed for certain types of ping pong ball, or 'matter', while allowing a special type of the smallest things, in essence a special type of the fizz from which matter is eventually built, to go faster than that maximum (possibly based on yet further restrictions to satisfy the whims of the Lord of the Ping-Pongoverse). Therefore Pyror, likewise, could allow certain things to travel faster than matter or other building blocks, or indeed light, if he so chose and according to house rules of his own definition. Ergo, tachyons can travel faster than light, because that is the will of Pyror.

Taking this Newfound knowledge as an opportunity to expand upon the options regarding your flying saucer's anti-gravity drive, let us assume that you have now upgraded it with a faster-than-light option. Therefore, to arrive at Zog before you left, the steps would be:

- Point your flying saucer at Zog.
- Set the automatic timer to one year and five minutes ago.
- Engage the tachyon drive.
- The tachyon drive takes over and converts you and your flying saucer to tachyons.
- Fly towards Zog faster than light with the anti-gravity drive on auto-space-pilot.
- Travel back in time one year and five minutes until the auto-space-pilot activates the timer.
- The tachyon reverse thrusters fire up.
- You and your flying saucer are converted back to regular matter.
- As regular matter can (by house rules) only travel at the speed of light, the conversion automatically slows you non-destructively to the speed of light.
- Engage the regular reverse thrusters, so the sub-light functions of the anti-gravity drive gradually slow you down to re-entry speed over the next year or so.
- Take back the controls and land safely on Zog five minutes before you left.

In fact it's fairly obvious when you think about it. Also, depending on the optional extras you can afford for your faster-than-light anti-gravity drive and/or the status of your relationship with Pyror, you could conceivably get converted back to regular matter that was *stationary*, thus avoiding the bit where you mess around for a year decelerating. The steps would then become:

- Decide where to go.
- Pray.
- Become tachyons.
- Revert to matter.
- Arrive fresh as a daisy in less than the time it takes to blink.

Rather ironically, once you've mastered that, you only really need the flying saucer as a convenient means of portable luggage storage and somewhere warm and dry in which to pray your way around the universe. Hurrah!

Chapter Nine

Bite Me!

20) I'm not sure what to have for tea tomorrow. Can you help?

Funnily enough, I'm very glad you just happened to ask that now, at what some might think of as a suitable spot for what may initially appear to be a somewhat tangential discussion. Talking of tangential discussions, if anyone is offended by the lack of an introductory artistic 'drop capital' as the first character in the body text of this chapter when compared to other chapters, I would have to repeat this chapter's title, 'Bite Me!' as a witty retort!

In fact what I'm going to do in this chapter is to suggest to you exactly what to eat, not just for tea tomorrow, but for every day for the rest of your life. By eating this you have the best chance of being healthy and happy, and therefore getting the most out of being alive. If you don't eat exactly what I tell you, you increase your risk of being erased from history ... Hang on ... My publisher's lawyer has just said that I need a disclaimer here ...

Disclaimer alert!

'Always consult your physician/GP before embarking on any health management programme. The direct orders included in this book for exactly what to eat should not be taken as direct orders for exactly what to eat, nor should any direct threats to have you removed from history for not complying with the suggestions, however foolish such resistance may appear to be given that Pyror himself suggested them to me when I was 'dead', be taken as an implication that you may be suffering from any form of cognitive folly, nor indeed as a direct or indirect threat that you personally may be removed from history or otherwise harmed in any way either physically or emotionally, nor should you infer that any changes to your level of health and/or happiness may arise from such an eating regime despite my direct and unequivocal assertion that *Pyror himself* told me to pass on this specific information verbatim et literatim insofar as an Outside-the-Box to Canine to Human translation would allow, which for those of you who speak less Latin than a GCSE Latin student translates as "word for word and letter for letter".'

End of disclaimer.

Apologies for the Latin rant by the way. Mad-B insisted. He's still traumatised by having been forced to attempt to do two years of Latin at school and allegedly spanked regularly for his failures, and really quite bitter that you can now get a Grade A Latin GCSE *without knowing a single word of Latin* because most of it is 'Roman Culture' written entirely in English: not just a *pass*, a *Grade A* ... I guess that given my Latin joke in Chapter Eight he was bound to crack sooner or later ... Sorry ...

So anyway, eat this or bite me!

Do *not* make any adjustments to the quantities or the ingredients under *any circumstances* or Pyror might immediately DESTROY THE UNIVERSE ... Nope ... Apparently my publisher's lawyer, not that some people might infer that there are likely to be many publishers' lawyers in the hologram, says that I need to provide various meal alternatives based on any specific requirements you may have, related to any or all of the following or any other special needs or desires or unusual lifestyles you may have, whether ongoing or transient, including age, gender, weight, allergies or intolerances, sexual preferences (WHAT …? I'm not doing THAT!), additional physical needs, cognitive impairments, ethnic backgrounds, challenging behaviours howsoever manifested, or of course belief or otherwise in Pyror and/or any other deity or deities.

The 'headline' quantities given are for a male who isn't on a diet. As we go, I'll also clearly show the amounts for women (roughly 20% fewer calories), men on diets (ditto), and women on diets (20% fewer still), although it's really quite simple as only *some* of the quantities need reducing but the balance *is* important for excellent nutrition.

I'll also show calories, contents, recipes, what to do for certain food allergies or intolerances, occasional cheaper or quicker alternatives, and the highlights of the most abundant vitamins and minerals in each item. For anyone who's interested, at the end of the chapter I'll probably have a quick bit on what the main features of each vitamin or mineral are, plus an absurdly nerdy breakdown of the composition of each item versus the Recommended Dietary Amount, aka Recommended Daily Allowance, aka RDA, ie the amount that avoids deficiency symptoms in roughly 98% of the population, which quite frankly is *not* perfect, but as the *large* majority of people are currently deficient in *something*, knowing that we exceed the RDA for *everything* is quite good for us!

Incidentally, on the subject of RDAs, ie the amount most

people need for 'good' health rather than just simple survival, they're prone to changing regularly. In the United Kingdom in the 1980s, you needed 30mg of Vitamin C. Then it went up to 60mg. Then at the time of writing it had gone back down to 40mg. Meanwhile, in the European Union it was 80mg. Sometimes it's all enough to make a dog take a break from wagging! Anyway, I'll stick with 60mg for now, not that you won't be getting nearly five times that, and not that I'm saying you'll get scurvy if you only have 59mg per day, because you won't! (Scurvy kicks in at around the under 10mg per day level if sustained for one to three months.)

If you're male, not on a diet, and you can afford to spend an 'average' amount per day on food, say two thirds of the cover price of a quality paperback work of quantum canine theology, for reasons to follow that I'll delve into ad nauseum (boom boom – it's a food-related anti-pun), eat this, but firstly Maxi's top tip: it's usually easier, more consistent and more accurate to weigh liquids on an electronic kitchen scale costing a tenner than to use a measuring jug (for liquids such as water and milk – denser liquids such as sulphuric acid and mercury are more awkward, though easy enough if you have a calculator and factor in the relative density aka specific gravity of course!). And while I think of it, don't panic; no one's actually going to destroy the universe just to stop you having the odd massive fry up if you feel the need, it's just best not to do it every day, but then you probably already now that.

So, if you're Mr Joe Average, eat this (and drink plenty of fluids, preferably ones that are good for you, eg clean bug-free water or, ironically, up to one bottle of red wine per week spread over several days, which often boosts good cholesterol:

that's *one* bottle of *red* wine per *week* spread over *several* days; <u>not</u> one bottle of red wine and two bottles of vodka!):

- Breakfast: 80g of porridge oats made with 370g of semi-skimmed milk (which is roughly 370ml if you're stuck with an old-fashioned jug).
- Mid-morning: three Brazil nuts plus some of the lunchtime vegetables (see below) if you have the munchies.
- Lunch: 200g of wheat-free rye bread (usually three to four slices depending on the shape of the loaf) with 40g of almond (or peanut) butter (but no real butter, not even delicious Wyke Farms butter which Mad-B still loves as a treat at Christmas even though they fired him for blithering incompetence, since when the value of the company has *increased* by around two million pounds, or butter substitutes, because to humans that would taste *weird* on rye bread!), plus circa 80g each of the following vegetables all RAW: red pepper, carrot, tomatoes and broccoli.
- Afternoon snack: an apple (from the Grand Appleverse if you like), an orange and a banana.
- High tea: a brisk 30 minute walk (or equivalent as a minimum if physically able).
- Dinner (with recipe to follow): Mad-B's delicious optionally spicy bean and lentil preparation with delicious wholegrain rice, with cayenne pepper, saffron, turmeric and/or soy sauce in moderation to taste if you like.
- Whenever suits: a tasty slurp of 5g of flax oil and a multivitamin tablet that includes multiminerals if you like, although many a dietician would tell you the tablet isn't necessary as

this eating plan is SO good. Personally, if I were a human, which I'm not, I'd have one, especially if you're taking any of the 'diet' options that follow or if you have any 'naughty' habits or feel stressed.

I very strongly suggest, if you're looking to lose weight, that when first switching to this eating plan (if you so choose) that you have at least a month in 'normal' mode while your body adjusts to healthy food, rather than diving straight into 'diet' mode. The lawyer wants the disclaimer mentioned again. He's not the most philanthropic human I've ever met. I certainly won't be affectionately humping *his* leg anytime soon (because I'm a *good* boy and don't do that sort of thing!).

In the rest of the chapter, we're going to cover exactly why each ingredient is included in the eating plan and what to do if you're allergic to any of those ingredients, or what to do if you want cheaper options, or what to do if you just don't like what's on offer (even if it might enhance, extend or even save your life to eat like Pyror asked me to suggest to you for your consideration).

I can cover the last point fairly easily. Let's take raw broccoli as the example of the thing that many people love to hate even the sound of. How about you? Does the very idea of it make you want to rush to the garden, eat some grass and regurgitate, or at least order a broccoli-free meatilicious pizza dripping with succulent oily cheese? If so, you could try one of two things: one, muse for a while that some people would be grateful for any kind of food, even raw broccoli ... Sorry ... The lawyer says it's a perfectly valid lifestyle choice not to eat raw broccoli and that I need to go on a 'sensitivity training' course – like *he* can talk! Or two, given that after a month I absolutely guarantee ... No, apparently I don't guarantee anything ... I wonder what a Quantum String Theoriverse with no lawyers would actually be like ... Heavenly I expect ... By pondering which point, apparently I intend no offence to lawyers or judges, or their long-suffering secretaries, paralegals or other staff or associates ... So, I therefore hope, although I can accept no liability for any pecuniary loss or emotional distress caused directly or indirectly as a result either to yourself or any other individual howsoever affected, that if to start with you just nibble a little bit of broccoli then immediately flush the taste and texture away with a vegetable you find more palatable, a firm tasty cherry tomato perhaps, that within a month or two you'll actually *like* raw broccoli and it will make you feel better and more righteous than anyone who *hasn't* fully embraced The Word of Dog. And of course you WILL be more righteous than any such person, whatever any lawyer might say!

a) **80g of porridge made with 370g of semi-skimmed milk (or vegan substitute if preferred, eg fortified soy milk, or if you just fancy a zero-cholesterol diet; yes, that's right, zero; not low, zero; so you won't need, for example, to eat anything specifically designed to block dietary cholesterol because you won't be eating any)**

The main highlights here, apart from it being a pleasantly sustaining breakfast, are that porridge contains a generous helping of soluble fibre which is usually great for starting to clean out any arteries that might be labouring under the weight of existing or self-generated cholesterol; meanwhile, the milk means you get a great start as far as calcium is concerned and all of the vitamin B12 you need (most vegan substitutes have added calcium and B12, but please check carefully if you're going vegan, or your nervous system will degenerate), so you won't have to eat any flesh (unless you're having a day off) which is great because if you're Mr Joe Average it means you'll live longer (unless you choose to believe in destiny and you're destined to be run over by a bus, eaten by a BEAR, killed by a falling bag of lentils, or something similar, in which case I guess it won't make much difference, although you might at least feel more fully alive in the meantime).

I'd strongly suggest that you don't add sugar, jam, honey or sweeteners because of the strain it puts on your body being forced to react to it, although if you must I'd do it in terms of weaning yourself off gradually over the first few weeks.

Let's look at what's in this breakfast, how to make it, and an alternative, although I do NOT intend to provide variations based on sexual preferences whatever the lawyer says! Remember, I'll just mention the exciting bits of the contents for now, but I'll include a full breakdown and analysis at the end of the chapter for anyone suitably curious.

- Calories: 469 for which you get over half of your daily calcium, all and more of your vitamin B12, and significant amounts of magnesium, zinc, iron, fibre, phosphorous, copper, manganese, folic acid and vitamins B1 and B5. Hurrah!
- Recipe: weigh 80g of 'proper' porridge oats such as Scott's

or Quakers, but by all means try others as long as they aren't powdered or processed, and 370g of semi-skimmed milk (total weight 450g) into a bowl big enough to allow the mixture to expand to about three times its size while cooking. Microwave to taste: probably about 3 minutes for runny to 4 minutes for gloopy in a domestic microwave oven, although by all means use a hob or other heating facility if you prefer.

- Allergies: try goat's milk, or fortified soy milk as mentioned in the subtitle.
- Alternatives: 100g of 'no added sugar' Alpen with 250g of semi-skimmed milk is a good alternative occasionally if you're in a rush, albeit that does have wheat and nuts which won't suit everyone, but most large supermarkets will have a similar low allergy alternative if you have a rummage.
- Women, or men on diets: have the same amount.
- Women on diets: have 50g porridge and 230g milk (293 calories) plus a multivitamin with multiminerals.

See how easy it is? Let's tuck in!

b) **Three Brazil nuts (plus some of the lunchtime vegetables if you have the munchies)**

Brazil nuts contain selenium, a valuable antioxidant which helps 'neutralise' harmful free radicals – reactive chemicals that can damage your body's cells and tissues. In fact, Brazil nuts contain a *lot* of selenium. So much so, and so much more than anything else (except maybe a chemical laboratory that has a jar of selenium in it), ie at least fifteen times as much per gram than anything else at all, that a theologian might attempt to argue that brazil nuts alone prove the existence of whatever deity the theologian prefers, although personally I would advise against that inference as the sole basis for a system of beliefs. In fact by far the easiest palatable and healthy way to get more than the 75mcg daily as now recommended by the UK government for men is to eat just three Brazil nuts. And by the way, that really is 75mcg per day, or 75 micrograms. Two whole grams would last you over 70 years. Hurrah for Brazil nuts! Eat them! Mind you, two grams of B12 would last you over five thousand years!

- Calories: 66 for which you get more than double the selenium you need and a useful amount of omega-6.
- Recipe: eat the Brazil nuts.
- Allergies: have a multivitamin with multiminerals.
- Alternatives: eat the nuts – they're good for you (unless you're allergic to them). A kilo of turkey giblets also contain quite a bit of selenium, ie roughly as much as three Brazil nuts, but many humans seem to prefer the Brazil nuts.
- Women, or men on diets: eat the nuts.
- Women on diets: eat two nuts (44 calories).

c) **40g of almond (or peanut) butter with circa 200g of wheat-free rye bread**

This delicious treat is what's going to sustain you through the afternoon and make sure you have enough good fats in your diet to carry essential fat-soluble vitamins around your body, plus to help build healthy cell walls. Do *not* avoid the fat from the almond butter: good fats in the right amounts (and this regime is actually plenty low enough in fat anyway) equals health, vitality and a longer life of better quality, subject to unexpected busses and the future consequences of any other pre-determined destiny in which you may wish to believe.

The lower amounts for women and dieters have been carefully chosen after due diligence and calculation, but *some* good fat is *essential* to health. Even my publisher's lawyer has been forced to concede that the life expectancy of an otherwise average individual, but whose overall body fat content is too low, is shorter than an otherwise equivalent person whose body fat content is within the healthy range (not that the lawyer is in that range because, like all lawyers, with the possible exception of any litigiously minded lawyers howsoever made party to the information herein contained, he's morbidly obese from eating port-basted swan stuffed with Armangnac-soaked foie gras every night). People 'without an ounce of fat on them' are *not* generally as healthy as they might be. They need more almond butter. Do not aspire to be like them. Do not aspire to size zero. Look down (with compassion) on such people, for they have not understood *The Word of Dog*. If necessary, add a bacon sandwich or a cheese pie to this way of living; get healthy and feel alive, but don't get too thin!

- Calories: 625, for which you get roughly two thirds of your iron and vitamin B1, a third of your calcium, copper and vitamin B3, half of your magnesium, vitamin B2, fibre, phosphorus and manganese, plus some valuable amounts of potassium and vitamin B6, and *all* of your vitamin E (another valuable antioxidant) and even a decent amount folic acid.

- Recipe: smear the nut butter on the bread and insert into your face in bite-sized chunks. Chew.
- Allergies: not many people with gluten issues are allergic or intolerant to rye bread. If you *are* you'll probably have to do what's best for you here. Gluten free rice bread perhaps, albeit it's expensive and not very nice? Of course if you're allergic to nuts *and* peanuts, you'll also have another level of problems to deal with, but unfortunately sometimes life means 'playing the hand you've been dealt'. Sorry that isn't more helpful.
- Alternatives: peanut butter is a lot cheaper but has less calcium, although you'll still get enough, so in this case feel free to choose, not that it isn't your choice anyway!
- Women, men on diets, and women on diets: have half the amounts of each (313 calories).

d) Lunchtime vegetation

Vitamins. Fibre. Immune system boosters. Wrinkle reducers. Vitality enhancers. What more could you possibly want?! The thing about vegetables is that many of them, especially these ones, contain traces of plant materials that do people a world of good (and probably dogs, although reliable data on canine vegetation-related nutrition is more limited), for example often in actively lowering cholesterol.

Genuine food allergies to fruit and veg are very rare indeed: you're far more likely to be run over by a fruit lorry than you are to be seriously harmed by eating its wares. You *might* go to the toilet vigorously and frequently when you first switch from any self-destruction in which some of you may have indulged to eating healthy vegetables, as your body grabs the chance to have a good clear out, but that's *not* an allergy. I did record one case while I was trialling this eating programme where someone had an all-over body rash for five days and frequent 'enthusiastic' bowel movements for two months, but that was rather extreme!

The amounts are the same for men and women whether dieting or not.

Raw red peppers. These are quite remarkable vegetables, mainly because a typical raw red pepper has at least *three times* the vitamin C of an orange, yet it is almost unbelievable (and disturbing) how many doctors tell their patients to eat an orange rather than half an average red pepper a day; half a pepper usually being near enough to the 80g or so that Pyror told me to suggest to you that you might like to think about eating (which for reasons that will unfold has a certain irony to it). Later on we're going to have an orange as well by the way but, when it comes to the unspoken micronutrients in fruit and veg, colourful variety really is the spice of life!

- Calories: 25, for which among other things you get almost twice your RDA of vitamin C.
- Recipe: chop out the bits that aren't red and eat the red bits!
- Alternatives: have a yellow or orange one if you must. If you're feeling skint, make a pepper last three or four days instead of two.

Raw tomatoes. Anti-aging, anti-cancer lycopene! Apart from tasting delicious (especially if you buy firm cherry tomatoes: in my experience as a tomato-eating dog, which I assure you I genuinely am, some of the others can be a bit wishy-washy, although it's really just a matter of personal taste and perhaps budget), a daily dose of tomatoes will also slash your chances of getting prostate or bowel cancer, which by no means just kills old people, as Mad-B's dead brother will tell you (slash being a non-specific 'number' that the lawyer insisted on, advising me to suggest that you should consult a suitable medical practitioner for an actual value, not that you're particularly likely to get a more specific answer if you do).

- Calories: 14 for which Mr Joe Average gets to live a bit longer in good health while looking younger. Tomatoes are an amusing little salad fruit/vegetable in that they aren't packed with anything obvious from the list of things with RDAs, but as mentioned they're loaded with lycopene which is *good* for you.
- Recipe: weigh, wash, gently pat dry and eat.
- Alternatives: if you're on a budget, buy cheaper tomatoes.

Raw green delicious yummy broccoli ...

Personally, I prefer to watch people eating pre-packaged pre-washed inorganic broccoli florets: the amount of bugs in the organic stuff can get pretty ridiculous at times, and from the look on most people's faces when they chow down on a large grub, I can only assume it may put them off, so inorganic broccoli may well be better from a long-term sustainability perspective. It's entirely your call though. Go organic if you like. Hurrah for broccoli!

- Calories: 27 for which you get lots of vitamin C, potassium and folates (folic acid), plus trace compounds that help with healthy hormone and enzyme production, thus again allegedly reducing the risks of some cancers.
- Recipe: weigh, wash, shake and pat until not too wet, eat.
- Alternatives: raw spinach makes for a delicious treat if you can't get broccoli. Also, if you've *actually* had a blood test showing hypothyroidism, I'd stick to spinach until you feel more lively, then maybe experiment. If you *think* you have hypothyroidism but you live on burgers, I'd go with the broccoli.

Raw carrots. OK, so carrots just *have* to be different, and the vitamin A (as beta carotene) with which they're packed is actually more accessible to the body after cooking helps break down the harder to digest fibrous materials wherein the beta carotene lurks. So, if you fancy cooking and cooling your carrots for lunch, be my guest, however personally I wouldn't recommend it, mainly because this eating plan is so high in vitamin A that it's actually not likely to be of any extra benefit messing around with cooking the carrots. In fact, too much vitamin A, like too much of many things, is distinctly bad for you. For example, you can get your full daily requirement of vitamin A if you eat one tenth of a *gram* of the liver of a Polar BEAR. Trying to eat a whole Polar BEAR's liver can kill you, especially if the Polar BEAR catches you doing it!

- Calories: 33 for which you get a really good jaw workout, plus of course the whole vitamin A thing.
- Recipe: scrub or peel, it's up to you, then eat the orange bits.
- Alternatives: have a pretentious purple carrot and eat the purple bits! If your teeth ache, or if you have no teeth, you could have carrot juice instead.

e) Afternoon fruit

Gosh. All this eating must be pretty tough on you humans. I think I'd find this many meals un**bear**able!

An orange. A welcome addition to any eating plan as, while we've already had plenty of vitamin C in our red pepper, vitamin C is one of a small handfuls of things where there isn't any known toxic dose, at least not if you stick below 30 grams a day, ie several hundreds of times the recommended daily dose wherever you happen to live, and 1,000 times the RDA of the 1980s before 'they' decided they'd been killing people with ignorance for decades, although apparently we can trust them now: personally, I'd prefer to trust Pyror and he certainly thinks oranges are a good idea. Plus, all of these fruits contain a good dose of fructose, which does sometimes mistakenly get demonised because it's a sugar, although it is actually **really useful in the mechanism for helping to build and maintain healthy cell walls**. So, munch away, but like everything do it in at least some degree of moderation.

- Calories: 61 for a biggish one, for which you get a healthier body.

- Recipe: peel the orange and eat the juicy flesh. You can eat the skin and bitter white pith as well if you're a Newfoundland with a penchant for oranges, but not many humans like those bits. Make sure you're authorised to help yourself from the fruit bowl though, or you might hear something like: 'Oh Maxi, you *are* a little scallywag … Aww … No, Daddy's not cross … Maxi want a tummy rub?'
- Alternatives: if you're on a budget, have a smaller orange.
- Women, men on diets, and women on diets: have a normal orange.

An apple. An apple a day keeps the doctor away, or so the saying used to go. Except of course these days it's five portions of fruit and veg, and it's not a doctor, it's an undertaker! And the portions are dumbed down to five instead of almost twice that because the government (or more accurately a largely faceless quango) decided they had no hope of getting people to eat a *good* amount of fruit and vegetables, so settled for a compromise that would at least help. This regime has roughly twelve portions a day (and in a good ratio and variety: no vegetables and twelve of the same fruits wouldn't necessarily be entirely sensible) based on the calculation of one portion being 80 grams of edible plant flesh, which itself is obviously yet another massive dumbing down in that it takes no account whatsoever of differing nutritional contents, for example gram for gram red peppers (for which rather ironically an 80 gram portion is eminently sensible!) have over 40 times the vitamin C of celery. The government must *really* hate you to treat you with such utter contempt. I recommend that you listen to Pyror's Little Helper instead!

- Calories: 62. Rats who eat six apples per day halve their risk of breast cancer. Statistics are weird, but probably not as weird as scientists who force-feed rats with apples. One apple is the most likely route to happiness (notwithstanding any sequelae of consuming a forbidden apple or apple-like fruit such as in the Garden of Eden in Creationist universes).
- Recipe: wash it if you like. If you're Newfoundland who's been scrumping from the lower branches of the trees in the garden, you might like to eat the core as well, but again that's less popular with humans.
- Alternatives: if you're on a budget, you can often get cheaper apples from a greengrocer, or in desperation you could have an extra banana instead. If money's not tight you could certainly use red berries or nectarines as substitutes if you like.
- Women, men on diets and women on diets: have an apple.

A banana. Lashings of potassium (do NOT eat lumps of potassium metal from your nearest industrial chemical processing facility or your face may explode, or would at best be permanently and hideously disfigured!) and tryptophan to allegedly help you sleep better, or so the rumours go, yet alas like so many things it's a bit of a fallacy. What they will do, however, is help you to stave off hunger and give you a slow enough release of energy to get you through to dinner. They do contain a *useful* amount of potassium compounds though, which are good for all sorts of bodily functions, such as heart function and conducting electrical impulses around the body, but with this eating programme you get more than double what you need without the banana. Likewise they do contain some tryptophan, but the porridge contains *far* more and will help you stay *awake* until lunchtime. I *suspect* that the mechanism involved is actually mood enhancing, so people sleep better when it's bedtime because they're happier in general, although of course very little about food and diet applies to everyone and the interactions are almost always complicated and on multiple levels. Nevertheless, because of the slow energy release and fairly low cost, I'd certainly get a banana down my neck if I were a human. Or indeed a dog. I like bananas. I, Max Merrybear, hereby affirm that I genuinely like everything in this eating programme.

- Calories: 118 for an average one for which you get a useful amount of vitamin B6 plus a bit more potassium.
- Recipe: I really hope you know it's best to peel it ... then eat the bit that isn't the peel ...
- Alternatives: if money's not a problem, have a fairly traded banana – they don't usually cost an awful lot more, you get to feel more altruistic, and it probably makes someone somewhere a bit happier, so ironically *they'll* probably sleep better as well.
- Women, or men on diets: have a banana.
- Women on diets: you can skip the banana if your body mass index is over 25 if you feel you must, although if you're obese it may be best to eat the banana anyway – don't go too crazy too quickly – it almost *never* works in the long run, and in (genuine) fact over 95% of people who start a 'diet' (rather than this, a healthy eating programme

for life) are actually heavier one year later, however fanatical their initial intent and fervent their self-belief.

f)   Healthy exercise

At the very least, if you're able-bodied, walk. You probably know the score, ie what the generic faceless quangos suggest for adults: thirty minutes of brisk walking five times a week. If it's not safe to go out, march up and down your living room. Or buy an exercise bike to sit on (and pedal) while you're watching TV. Or a 'get fit' games console or workout program on DVD for example. If you can't do any of those because of cash, medical reasons or disabilities, get your GP to refer you to a personal trainer or occupational therapist who can give you specific advice. If you can't do any of those because you're too busy, change your lifestyle until you can (unless you're happy being miserable – no one has the right to force you to exercise as long as you 'own' the consequences). But if you can pick up this book, you can probably exercise. If someone else is reading this book to you because you *can't* pick it up, I *might* just accept a PE note from your reader.

Actually, where is Mad-B when you need him ...? All this talk of exercise has made me fancy a 30 minute walkies ... Oops, I think I left him in the garage making Perspex cubes ... He is probably NOT going to be happy. Maybe I *am* a little scallywag after all!

g)   Mad-B's delicious optionally spicy bean and lentil preparation with delicious wholegrain rice ... Mmm ... Delicious ...

For those of you eating as a family, or anyone prone to indigestion or other challenging digestive behaviours, it might be best to go easy on the spices when making this thoroughly delicious meal that Mad-B cooks me and Jazz occasionally, especially if we've gorged on pheasant stuffed with Stilton and are feeling a little bunged up: you can always put a jar of cayenne pepper on the table for those who like it a bit spicier, or even better a *little* Hillboy Chilli Sauce available at the time of writing from www.gurkhafinefoods.co.uk. It's *hot*. Or just have a side plate with a few extra chillies to nibble. Whatever suits.

So. Lashings of protein, slow releasing carbohydrates and plenty of dietary fibre to help flush toxins where they belong: out of your body rather than festering in your intestines. And while we're telling the truth, the whole truth and nothing but the truth, what healthy-bowelled dog or human doesn't feel great after visiting the toilet or garden, or in Jazz's case any of the beaches at Studland in Dorset, England, where he has 'eliminated stool' in the sea every single time he's ever been there. I have a absolutely no idea what it is about Studland that sets him off, and he refuses to discuss it with me other than to say 'when you've gotta go, you've gotta go,' and he's never done it at any other seaside resort, but as soon as he gets to Studland he runs into the sea, just out of paddling range of Mad-B, squats and evacuates, however much Mad-B pleads or screams at him to stop. As Jazz is Glaswegian by birth I can see no obvious way in which this particular foible may be a repressed puppyhood memory, ie he had probably never seen a beach before I met him, so it doesn't appear to relate to any earlier trauma he may have suffered.

Still. Protein. The thing about proteins is that they're made up of amino acids, essentially one of the building blocks of life itself. In fact it seems likely that Pyror has a separate subsection of house rules especially for amino acids, just like you, Pingpongor, could, if you so chose, have a separate subsection of house rules for what happened if you got a clump of say exactly four ping pong balls touching, thus allowing your ping pong balls to stay together while moving on to even greater things. Mind you, Pyror would also need to have different subsections of rules for how the assorted atoms that make up the amino acids behave so that they can form molecules involved in the biochemical reactions based on another set of rules, all subservient to the rules about how many classical atomic particles make up what sorts of elements, and further subservient to the way quantum energy strings behave in order to form what appears to be the solid matter of classical particles so, as you can see, in some ways the universe is actually fairly complicated: I'm not sure that I'd like to try to make up the rules and still expect it to be a *good* universe!

Anyway, it's time for dinner, and while there are lots of different amino acids, some can't be made in the human body, so for good health it's essential to get enough of them in your diet. How many, it's hard to say, but for most people it's either eight, nine or ten, which is interesting: given the right ingredients, your body knows exactly how to make the other dozen or so amino acids it needs (dietary science really IS still in a state of constant flux!), but food scientists are sometimes so befuddled (relative to your body I mean) that they don't even know how many can only be acquired by eating them, let alone exactly how to make them, but they'll still have heated arguments and fights with sharp sticks about how many are actually essential in the diet, very much like astronomers still fighting over whether there are eight, nine or ten planets in our solar system (Pluto and Charon being the sources of the astronomical chagrin)!

So. Beans and lentils. Eat both together as suggested and you'll get all the high quality balanced amino acids you need for a protein-packed body. Yum yum. And with all that fibre you'll go a long way to avoiding bowel cancer. And beans actually count as (a maximum of) one of your (simplistic) five (or ten) portions of fruit and vegetables a day, with the tomatoes indicated below adding another two and a half portions. And the rice is full of good starchy carbohydrates to sustain you through the evening, plus combined with the protein from the beans and lentils will give your body all it needs to replenish itself after your healthy exercise. If you want to aim to be super-healthy, by all means add some spinach. Hurrah for spinach!

- Ingredients (approximate weights): 200g of tinned beans of your choice, drained if appropriate, 71g lentils (one seventh of a 500g bag), half a 400g tin of tomatoes, 125g brown rice, and to taste your choice of spices and 'free' spinach.
- Calories: 943 for which you get everything and more that you need that you haven't already had.

- Recipe: ah, there are oh so many ways I've seen Mad-B experimenting with the preparation of this, so I'd suggest that you find the one that's right for you, but here are some pointers with a recipe to follow that balances quick with cheap and easy. If you're using cheap brown rice (which is a good idea for most people, because it's cheap and usually just as nutritious as other sorts!), it'll probably want between 25 and 35 minutes of simmering depending on your tastes (unless, allegedly and for reasons that will eventually become clearer, you're a Ninja and prefer to eat it raw). If you use Uncle Ben's Parboiled Wholegrain Rice it's 11 minutes, so better if you're in a rush, but up to four times the price, but you can use the absorption method to soak up all the liquid, so you can flavour it with optional soy sauce more easily (ideally delicious 'Tamari', often found in the gluten-free sections of supermarkets), or even saffron for a treat. Red lentils need little more than a couple of minutes boiling before you can add all the other ingredients. You can adapt the recipe to use just a single bowl and spoon if you believe the act of washing up to be a crime against Belphegor (one of the seven princes of Hell, who seduced people by tempting them with laziness, often in the form of get-rich-quick schemes, and was sometimes worshipped in the form of a phallus: the mind really does boggle sometimes). You can fry a spoonful of curry powder if you like, ideally in a little spray-on sunflower oil, but not flax oil as that gets 'damaged' more easily by frying. You can just stir in some curry powder if you like if that's easier. It depends to some extent how much you enjoy fine dining and how much you want something fairly easy to prepare.

Here's one cheap, quick and easy method: by all means adapt it if you like. 1) In one saucepan, boil the rice and simmer to taste. 2) About ten minutes before your rice is due to be ready, in a second saucepan, boil the red lentils for a couple of minutes then drain, although by all means start earlier and give them a few minutes longer if you prefer. 3) In the second saucepan, add everything except the rice and simmer for a few minutes. 4) Drink the flax oil then serve Mad-B's delicious optionally spicy bean and lentil preparation with delicious wholegrain rice. Simple!

- Alternatives: if you're on a budget by all means have value beans, but conversely if you find a tin of curried beans you like that aren't too high in salt by all means go for it; at the time of writing Morrisons had a suitable and tasty offering. Just occasionally, if you really are in a genuine hurry, you could do worse than have a big plate full of microwaved Heinz baked beans (or similar) on plenty of toast with a bit of cheese topping (Wyke Farms Extra Mature cheddar being de rigueur as you may already have inferred), but very easy on the butter!
- Women, or men on diets: have 50g of lentils and 100g rice, otherwise the same.
- Women on diets: have 50g of lentils and 75g rice, otherwise the same.

h)   5ml flax oil

This stuff, also known as linseed oil, but NOT to be confused with the gone-off stuff you put on cricket bats to keep them supple and crack-free, is the brazil nut of the oil world: one 5ml teaspoon per day gives you all the Omega 3 you need, which is quite handy because it's essential for good brain function, building healthily flexible cell walls, and is highly likely to help to a greater or lesser extent if you have achy joints.

A good daily dose of Omega 3 will discourage your body from using bad fats to maintain and replace cells: too much saturated fat and not enough Omega 3 equals more rigid cell walls which then struggle to let nutrients in and toxins out efficiently, which puts your body under strain, which reduces your life expectancy and certainly has the potential to make you more lethargic until your untimely death – not exactly an inspirational way to live! And the extra great thing about flax oil, while it may not exactly seem dirt cheap to buy a bottle, is that even if you're the only one drinking it it'll stay fresh and effective until you finish a typical 240ml/260ml bottle: around eight weeks of full efficacy if you buy it in a dark bottle and keep it in the fridge. And after a month you may even start to enjoy the taste ... Hurrah for flax oil!

It would be remiss of me not to mention that at the time of *writing* all of the major supermarket chains have stopped stocking pure unblended flax oil, although hopefully by the time of *reading* that will have changed again. If not, you can get it by mail order from Granovita, although you may like to form a local '*The Word of Dog Flax Oil Syndicate*' to keep postage costs down: don't buy more than eight weeks worth at a time though, as the quality *does* deteriorate and eventually it'll go rancid and fishy. Or you could decide to eat an equivalent amount of flax seeds/linseeds, preferably ground and thoroughly chewed, as like sweetcorn they do pass through undigested given half a chance. If would, however, also be remiss of me not to mention that ground flax seeds often stick to people's and dogs' teeth in a somewhat less than enjoyable way.

- Calories: 41 for which you get all of your omega-3 which means healthier cells, and possibly better skin and more brain power.
- Recipe: pour some into a measuring spoon and savour the delicious taste. By all means have 10ml if you like, especially if your joints ache: it'll probably be calories well spent.
- Alternatives: linseeds as mentioned, but chew them well or grind them.
- Women, men on diets and women on diets: have the same.

i)   **A multivitamin with multiminerals**

Let's have a very quick look at what this 'stuff' is. After all, if we're going to shove it down our necks and digest it, it's probably worth knowing a bit about it ...

First of all, you can get a perfectly adequate product very cheaply from most major supermarkets, eg 'Tesco's Complete Multivitamins + Multiminerals'. Centrum Performance is a good all-rounder that some people will doubtlessly think is more balanced if they think supermarket cheapies are too generic, although given that this diet gives almost everyone everything they need (ie unless perhaps you have adverse medical conditions or a gastric bypass for example), I would regard the difference as negligible. By all means form your own opinion though. Nothing about *The Word of Dog* should be regarded as a quick fix that allows you to abdicate responsibility for your own opinions. Where's the fun in that?! And by all means have an extra vitamin C tablet every day if you like. The simple fact is that people who *do* have a 500mg vitamin C tablet every day, while doctors usually insist that it isn't necessary, allegedly get on average about 50% less common colds than people who *don't* take a tablet, and I understand, based on hearsay and Chinese whispers, that the colds don't hang around as long on average when they do appear. By all means trust your GP instead though if you trust them more than a big furry puppy or your own intuition.

So, if you want my *opinion*, have a complete multivitamin with multiminerals and a 500mg vitamin C tablet as well. More often than not they do more good than harm. But please do make up your own mind: better to be an independent BEAR than a herded sheep!

What does it all do? Well for a detailed discussion you could probably do worse than read *Nutrition for Dummies*, although that publication does make some people angry. Or Wikipedia. Everyone loves Wikipedia, right? No? Maybe? The point is I honestly don't think *The Word of Dog* is the right forum for an in-depth heated debate on which nutrients are involved in which biological processes, especially as nutritional information is prone to changing substantially from year to year whereas *The Word of Dog* should last for millennia. But, I did say that 'at the end of the chapter I'll probably have a quick bit on what the main features of each vitamin or mineral are'. Therefore:

- http://www.nhs.uk/conditions/vitamins-minerals

Yes, that's right, despite my earlier seditious references to 'they' and Mad-B now screaming in my ear about conspiracy theory and state repression, I'm seriously suggesting that you get your further information (if required) from a UK government website (as long as you can make your own mind up about which bits to believe). Go to a USA one if you must, however the US government recommends a lot more for some things, such as calcium, but my inference is that they appear to do it solely as an attempt at compensatory or prophylactic medication because of their obesity problem and the consequent leaching of calcium from the body by the much more acidic diet of an average American. So, the UK Food Standards Agency website: sensible practical information in an easy to read format. Sorry Mad-B, sometimes not even governments and quangos are beyond redemption in Pyror's eyes!

For any **nerds** who feel slightly cheated, here instead is the funniest ever pure maths **NERD** joke, and if you're a proper nerd there really is a genuine risk that you'll actually die laughing at this one. I trust that you won't mind me saying I am unable to accept any responsibility whatsoever for any injuries, fatal or otherwise, that this joke may cause.

WARNING: SEVERE NERD 'HUMOUR'!

$2^{GOOGOLPLEX}$ mathematicians walk into a bar. The first mathematician says to the barman, 'I'll have one over zero factorial imperial pints of beer please.'

The barman replies, 'I think I see where this is going,' and pours the mathematician one of Her Britannic Majesty's imperial pints together with one unashamedly metric litre of beer, filling the glasses.

The mathematician thanks the barman saying, 'Aha, I note that whilst the prevailing weights and measures legislation allows the head of the beer to count as part of the beverage, you have in fact kindly given us approximately 1.5% extra as gratuitous head!'

'Ha ha ha,' they giggle in peaceful unison, embracing the similarities in their quasi-parallel universes, swapping flowers, and sharing a non-abusive group hug with some double slit experimenters and smelly hippies.

I'll explain it later ... OK, it's later ...

Firstly, saying '$2^{GOOGOLPLEX}$ mathematicians' is the modern alternative to the now risibly outmoded 'infinite number of mathematicians'. That bit's simple enough. Next, one over zero factorial is the first part in the infinite (sic) series used to calculate e: one of the most useful constants in mathematics, perhaps second only to good old π. According to Wikipedia: 'The mathematical constant e is the unique real number such that the value of the derivative (slope of the tangent line) of the function $f(x) = e^x$ at the point $x = 0$ is equal to 1.' That certainly sounds useful to me (and actually, perhaps ironically, it is, and indeed we ourselves used it in some of the nerdisms in Chapter Two, such as =EXP(4*LN(2)) where *e* is used behind the scenes as it were in the logarithmic calculations). One over zero factorial equals 1 (by virtue of an anomalous mathematical diktat: technically it would normally be infinity (sic))! Extrapolating the series, one over one factorial also (correctly) equals 1. One over two factorial is 1/2. Then 1/6. Then 1/24. And so on. Add them all up and you get *e*: circa 2.718282. One litre of beer is 1.759753 pints, so add another pint and you get 2.759753 pints. Divide *e* by that and you get 0.984973. Express that as a percentage rounded to one decimal place and you get 98.5%, which is the percentage of the volume occupied by the liquid beer. Ergo, the 1.5% that remains is the head. The prevailing weights and measures legislation allows this to be counted as part of the beverage, but the barman has chosen not to charge for it, hence gratuitously giving the mathematician head. Finally, the reference in the joke to double slit experimenters and smelly hippies is a cliquey but non-assholoid nerd loop-link to *The Word of Dog* which itself also both contains and explains the joke, and therefore one might say arguably adds weight to the theory that the universe may be a pan-dimensional hyperbolic Möbius ouroboros. The joke is therefore, and on many levels, the most hilarious in the universe ever (for nerds).

Back to normal, thank goodness!

If you want to give a nerd an atomic wedgie for the above, I don't think Pyror will erase you from history for it … If you aren't familiar with atomic wedgies, it's certainly worth typing the same into Google and also clicking 'Images' while you're busy looking up more information about nutrients …

j) Summary

So, that's it. You now know what to have for tea tomorrow and how to eat every day thereafter if you want to live a long and happy life (accidents, disasters, assholes and plain bad luck permitting). Me, I think I'll keep asking for the odd bit of chicken liver pâté occasionally: I certainly like some of the better quality dog foods, but sometimes variety is the spice of life and I know Pyror wants me to enjoy being alive, although I do sometimes think Pyror wishes he could make a house rule that allowed dogs and people to enjoy chocolate and other food or treats even more without it becoming a dangerous burden to some, but then as you, Pingpongor, already know, sometimes it's just not that easy being god ...

And of course, as mentioned earlier, one of the great things about this way of eating is that it costs no more than a couple of lattes in a coffee shop chain; significantly less if you go for the cheaper options to scrimp and save, more if you have your manservant pop down to Fortnum and Mason's in your Rolls-Royce to buy the most expensive or pretentious organic artisan options he can find.

Right. For those of you with no particular further interest in the nutrition of what you just ate and are happy to know that, as long as you're eating normally, even without a tablet you've had over 100% of your daily vitamin and mineral needs for *excellent* health (on average: subject of course to food being a natural product which does go off if you keep it for too long, and the precise composition of which is prone to seasonal, storage and other variations; and different people having different absorption abilities or challenges), with twice the protein you need, low saturated fat, excellent omega 3, and superb carbohydrates, you can skip ahead to the next chapter if you like, secure in the knowledge that until today 95+% of the population were deficient in *something* but that they no longer need to be. And the beauty is that even if you're a woman on a diet I've balanced the reductions so that you still get superb nutrition with 130% of everything you need with just a cheap multivitamin and multimineral tablet.

For the rest, as promised, there will follow some spreadsheets with the complete breakdown of the various vitamins, minerals, proteins, carbohydrates and good balance of fats, also including the ten amino acids that cover all the essentials, plus the typical cost at the time of writing (which will obviously vary a bit depending on what fruit and veg is in season).

For those of you for whom the font is a bit small, you could try one of the following:

- Visit www.maxmerrybear.com where I shall do my best to post the info in a more accessible format once my website's up and running, which it should be by the time you read this, unless maybe you've been given an advance copy or early first edition. By all means check out the merchandising section while you're there if I will have one by then. Buy a T-shirt if you like. Or maybe even a flying saucer. Where possible, the merchandise is probably all going to be ethically sourced and with a bit of luck the chances are most of the proceeds will go to providing the workers with a decent living wage and to other good causes, although obviously I'll need to sell quite a few books before I can buy a factory!

- Or buy a magnifying glass.
- Or ask someone to read it out to you.
- Or just trust me; after all, I am a dog!

k) **Nerd alert!**

Here are the spreadsheets with the displayed values appropriately rounded (so the total may look like it disagrees slightly with the sum of the displayed figures in places because of the unseen lower-order digits). Ignore these tables if you're normal.

As it might look like too much padding here, I'll probably put the detailed breakdowns for women, men on diets, and women on diets on the website if I have one, but in summary the rounded headline figures are:

	Men	Women and Dieting Men	Women on Diets
Calories	2471	1991	1596
Protein	90g	74g	62g
Fat	58g	44g	35g
Carbs	412g	338g	266g
Minimum % of RDA	114	102	Have a multivitamin
% RDA with multivitamin	179	152	130
Fibre	65g	53g	46g

Nutritional Breakdown – Table One of Five for Men not on Diets – Various Bits

Portions of fruit/veg	12	B'kfast	469				Costs
Calorie breakdown		Lunch	724				Monthly
From Protein	14%	Snacks	294				141.37
From Fat	20%	Dinner	943				Daily
From Carbs	66%	Oil	41				4.56

	1990	2471	90	58	412	2.3	4.56	3.1
	Weight	Calories	Protein	Fat	Carbs	Sodium	Cost	Omega 3
Oats	80	284	8.8	6.4	48.0	0.0	13	0.1
Milk	370	185	13.3	6.7	17.8	0.2	30	0.0
Brazils	10	66	1.4	6.6	1.2	0.0	9	0.0
Rye bread	200	372	11.6	2.6	75.4	1.2	38	0.0
Almond Butter	40	253	6.0	23.6	8.5	0.0	51	0.2
Toms	75	14	0.7	0.2	2.9	0.0	25	0.0
Broccoli	80	27	2.3	0.3	5.3	0.0	16	0.0
Carrot	80	33	0.7	0.2	7.7	0.1	8	0.0
Pepper	80	25	0.8	0.2	4.8	0.0	41	0.0
Banana	118	105	1.3	0.4	27.0	0.0	23	0.0
Apple	120	62	0.3	0.2	16.6	0.0	25	0.0
Orange	130	61	1.2	0.2	15.3	0.0	34	0.0
Red Lentils	71	246	17.8	1.6	42.3	0.0	13	0.2
Brown Rice	125	463	9.9	3.7	96.6	0.0	16	0.1
Tinned Tomatoes	192	61	3.1	0.1	14.0	0.0	25	0.0
Baked Beans	207	164	9.7	0.4	26.7	0.5	27	0.1
Flax Oil	5	41	0.0	4.3	0.6	0.0	10	2.3
Soy (Tamari)	5	3	0.5	0.0	0.3	0.3	9	0.0
Water	1500	0	0.0	0.0	0.0	0.0	42	0.0
Cayenne	2	6	0.2	0.3	1.1	0.0	3	0.0

Actually, as these really are quite dull, I'm going to shrink the rest of them a bit to avoid offending normal people …

Chapter Ten

MAQs II

Hey, that's my name: MAQs II … Maqs II … Max-ii … Maxi … Get it? BEAR. Oops, that's the third time I've said 'BEAR' (oops) for no apparent reason other than a spot of random hilarity, ie discounting any valid contextual uses of 'BEAR' (oops), '**bear**' (oops), '*bear*' (oops) or just plain 'bear' (oops).

So anyway, in this chapter I'll be proposing answers for your consideration to More Asked Questions II, ie another bunch of questions, this time eminently more sensible but still without implying any specific hierarchy to their posing, even if the answers are now building on our knowledge in a way that makes them ever more close to the most important question ever asked (in a later chapter) … Although, now that I mention it Question 22 is a little, ahem, 'out there', what with the whole Tooth Fairy thing … And Question 24, 'What?!', probably errs on the side of succinct … But on the whole they're mostly at least sane … Mostly …

21)   How guilty should I feel about 'Original Sin'?

A lot depends on your own personal beliefs and your own interpretation of whatever universe you happen to inhabit, plus the history, as perceived by you, of the said universe. The concise answer is therefore going to be either: a) not at all, b) a bit, or c) incredibly guilty and barely able to survive the shame of your repugnant and disgusting actions. Let us discuss.

Wordy sentence alert.

Firstly, for the purposes of this discussion and whilst not wishing to imply that you should alter any belief system you hold, unless of course you're searching for a new belief system, in which case feel free to believe fervently in whatever system you settle upon, whether proposed or implied in *The Word of Dog* or otherwise, I shall define 'Original Sin' as where a person or other animal living today, whether male, female, transgender, hermaphrodite, or castrated or otherwise neutered, is responsible for the sins or other misdeeds of their forebears, whether or not those misdeeds be a complete or only partial abomination unto their chosen deity or deities if any.

Back to normal.

By all means make the definition more specific (or even more woolly) if you like. For example a male Creationist might like to say, 'Original Sin is the fault of Woman for tempting Adam in the Garden of Eden, so we are all responsible for that guilt and shame, and must all repent every day.' However, for the purposes of this discussion, I'll go with my somewhat more generalised and inclusive definition if that's OK.

Even wordier sentence alert.

I suppose, if it's what you want to believe, no one has the right to tell you not to feel racked with shame at the disgusting behaviour of your ancestors, whether that be scrumping a particularly succulent forbidden apple to give to a friend, or indeed a slightly mistranslated peach or apricot as it's more likely to have been if the Garden of Eden were located in the Middle East circa six thousand years ago in most of the universes I know about, not that I'm saying it isn't perfectly reasonable to believe that Eden, if you choose to believe it existed along the lines of the Old Testament accounts, might have had a solitary apple tree or perhaps alternatively have been located somewhere near an orchard close to Glastonbury in Somerset, England, given that that's exactly where King Arthur's Camelot was located (his birthplace being Cornwall) and in a sealed cave deep beneath the Tor is, according to local legend, buried the Holy Grail, which may be useful information to Mad-B's son Tom, given that during a drunken betting session Tom wagered with a friend that he will find the Holy Grail before he reaches the age of 35 (human) years, with the loser agreeing recompense by consuming one imperial pound of horse faeces, but I digress; or whether it be the sin of founding a high street bank on the back of blood money earned from the slave trade, not that Mad-B is slightly bitter and twisted about his time in retail banking, but he insisted I mentioned that, along with one or two other noteworthy conditions about which we shall learn more in due course, in exchange for services rendered by his opposable thumbs – a reasonable compromise given the amount of time I would otherwise have taken to type this using chopsticks held between my 'fingers' to poke the keyboard, especially given my advancing years and breed-appropriate size 14 partially-webbed paws.

Back to normal.

Anyway, much as in the same way that you might be reincarnated as the ghost of a fictional television character (and remember we *did* actually establish a mechanism, however unlikely, whereby that could theoretically happen), I see no reason why the *burden* of sin, which after all is a thought stored in the brain using some sort of 'matter', however ethereal or quantum-based, could not undergo a similar type of energy transfer. And given that some people do feel burdened by the sins of the past, most Germans for example, I think we have to allow for the possibility of some sort of weird goings-on affecting at least some people, although similarly if you want to believe they're just internally burdening themselves unnecessarily with no actual transfer of external energy that's also fine by me, and in many or most cases probably more likely. But equally it doesn't seem to be *impossible* to transfer the *energy* of sin, and it doesn't seem to break any house rules.

Another wordy sentence alert.

Likewise, if you feel a *bit* guilty, the two main possibilities are that perhaps there was a *little* bit of energy transferred over the ages and into your body, or your general aura-based spatial presence: your own individual Oort cloud if you will, ie the region of the omniverse you may think of as your personal space, in other words the variable bit surrounding your material presence in which you feel uncomfortable on noticing unwanted intruders, whether that be wishing someone didn't stand quite so close to you, or indeed wishing that they went and stood on another planet altogether, hence variable based on the level of mutual attraction or repulsion between you, not that the region has a scientific name as such, or not that I'm able to discern, so henceforth I submit that it should be called the Maxisphere, as in the phrases, 'Hey, dude, get the heck out of my Maxisphere,' or 'Oh baby, I want your Maxisphere so much; I promise I'll still respect you in the morning …', or of course perhaps no actual energy whatsoever was transferred over the ages and into your Maxisphere but you still feel a *bit* guilty because you think you ought to.

Back to normal.

Finally, if you feel no guilt whatsoever, there are also two possibilities: one, that you feel no guilt whatsoever because no ancestral guilt energy was ever transferred into you or your Maxisphere (whatever may or may not have happened to anyone else), so you personally genuinely aren't guilty by association of anything whatsoever and also you don't think you should feel guilty about things over which you had no control, and two, guilt energy was transferred into you but is currently wasted on you because you seem to be oblivious to it, in which case you really ought to get down on your knees and beg for forgiveness, you filthy miserable original sinner. I imagine, for most of you, that it's the former.

What do I think? I think that if Pyror wants to win the Best Quantum String Theoriverse competition, and I think he does, then sooner or later he'll find a way to stop people and dogs feeling guilty and burdened by the sins of their ancestors, even if those ancestors have been a bit naughty. Pyror wants everyone to be happy. Even Roman Catholics.

22) I want to travel back through time and kill Temüjin before he took the name Genghis Khan and murdered a large chunk of humanity. How do I go about it? Please give your answer with reference to Doctor Who, Father Christmas and the Tooth Fairy.

Are you on drugs? If not you probably ought to be; and please allow me to reiterate that as previously mentioned any such personal slights or libellous remarks henceforth and hencebackwards always refer to the person who originally asked the question, for it would be an unwise hound indeed who tangled with or upset you, Pingpongor!

I suppose the first question is, 'Is time travel possible; specifically, in this case, time travel back into the past?'

Very long answer alert!

Technically, I have already hinted at a possible mechanism, given that if your wattle and daub flying saucer were fitted with a Tachyon Drive and programmed appropriately, you could set off for Zog at lunchtime and arrive in time for breakfast the same day, although this should not be confused with the popular yet often spurious tabloid allegations that a typical university student likes to eat breakfast at around 4.00 pm. If I'm honest, I can foresee a few problems with getting such a mechanism to work in practical terms, at least based on our current level of knowledge of tachyons, and certainly in terms of whether or not anyone or anything would be able to switch you back into a life form based on classical matter after you'd been converted into a stream of tachyons, but the theory's perfectly sound.

There is, however, one major obstacle: until around about now, ie once you've finished reading this answer, the concept of time itself has generally been somewhat poorly understood, and indeed you yourself may once have heard someone say, 'What if you go back in time and kill your grandfather,' or 'I'm right brainy, me; I've worked out that time travel isn't possible otherwise people would come back from the future and visit us, and buy a lottery ticket while they were at it.'

Here, naturally, is what would actually happen. But don't take my word for it, let's ask you, the mighty Pingpongor!

Let us just remind ourselves briefly, given that there are a few more time units in Pyror's full and unabridged Quantum String Theoriverse, and not implying that you, the mighty Pingpongor, couldn't have a lot more 'time' units in your Ping-Pongoverse if you wanted to, that for the purposes of this discussion you have three boxes, each representing one time unit of a single universe that could, if you so chose, have a quasi-parallel universe next to it, but at the moment it hasn't, so there's only one universe in your omniverse and only three boxes.

So. Three boxes. And an adequate number of ping pong balls. All obeying your house rules.

Let us, ie you, take a ping pong ball out of the final box (ie Chum-2's go: the last unit of time), call the ping pong ball 'Percy' and place Percy in a different square (or cube if you're playing in three dimensions) in the first box (ie the beginning of time when you, Pingpongor, created the universe). We can therefore say that Percy the Ping Pong Ball has travelled through time and space because in this case you, Pingpongor, willed it so. Easy peasy.

Now let's complicate things a bit.

Let us assume that you, Pingpongor, placed the Future Percy in a square next to the Original Percy, although for simplicity's sake we will, for now at least because it makes little difference to any paradoxes that may arise, assume that Percy was not previously involved in any ongoing creation or destruction of matter rules that you may have decreed, ie Percy was present for all three seconds of the Ping-Pongoverse, rather than being 'born' halfway through time.

This now means that both Original Percy and Future Percy would be standing next to each other and would be able to 'see' each other. If you equate this to how you'd feel if 'you' from two seconds in the future suddenly appeared from 'thin air' right before your (own) very eyes, I think we could safely say it might well drive both Percys insane (because it would drive Original Percy insane to see himself appear out of thin air, so he'd probably still be insane two seconds later when he travelled back in time to see himself). And given that only a benign deity, or at least one who is trying his or her best to be benign, has any realistic hope of winning a Best Ping-Pongoverse or a Best Quantum String Theoriverse competition (unless of course the judges are evil and/or insane, which personally I rather hope they aren't!), there are only two sensible courses of action open to you.

One, you could just take the easy option of making it against the rules for anyone, even yourself, to move ping pong balls between boxes. Time travel would therefore be, barring any cheating on the part of yourself, your chums, or an earthly version of the drunken uncle who po(o)pped in to see Pyror earlier, impossible.

Two, you could allow time travel but then immediately resolve any paradoxes according to your own house rules so that no one and/or no thing went insane.

Wordy sentence with big brackets alert.

Therefore, in essence, if your ping pong ball from the third box got moved back to the first box, whether by the hand of yourself, Pingpongor (or because a ping pong ball had managed to build a wattle and daub flying saucer with the tachyon drive needed for backwards time travel, or indeed an anti-tachyon drive which may in theory achieve the same results, and in this case I do of course mean anti-tachyon in the sense of moving at a negative velocity rather than having an inverted particle spin, negative velocity of course being impossible just like moving faster than light in the first place, which is what tachyons actually do in theory at least, or indeed impossible just like having imaginary numbers, for example the square root of minus one, which is 'impossible' but also rather strangely absolutely essential for electricity to function, but I digress), the paradox of having two identical ping pong balls would have to resolve appropriately.

Back to normal.

This would most easily be achieved by deleting the one which violated the most house rules, or in the case of a highly unlikely draw, tossing a coin to decide which one to keep. So while it would almost always be the time traveller who broke the rules (by appearing inappropriately in what should have been empty space), ironically it does actually provide a mechanism for a person in the Quantum String Theoriverse, other than through substance misuse, thinking 'How the blinking heck did I get here?'

If the impostor *was* successful in replacing the existing version of itself, all that would then need to happen is that every affected box (or time unit) would need to resolve any conflicts according to the house rules, whether your rules or Pyror's. Therefore, in this scenario, if you travel back in time successfully the old you is deleted, and before you have realised anything was going on you're in a particular place and probably with no real idea how you got there (because the memory of travelling back would usually need to be erased as well, along with the actual time machine, although it's just possible that there could in theory be some case-by-case exceptions).

The consequence of this, which is decidedly ironic, is that it would look to an observer as if nothing particularly odd was happening. This is because if a person travelled back in time, the earlier or later time units of the Quantum String Theoriverse would have to resolve such that the person was born, lived and died earlier in history (relative to what would essentially be an earlier run of the Quantum String Theoriverse game), rather than just popping in for a few hours like Doctor Who. This is true *even if* the person travelled back to *before* the original version of themselves was born, which even more ironically doesn't totally preclude them from living twice, as long as any paradoxes resolved to form a coherent Quantum String Theoriverse, and technically provides an alternative mechanism for what we could loosely term reincarnation, albeit the second life would in this case occur first and might therefore better be referred to as pre-incarnation!

Allow me, if you will, to clarify: if *you* travel back 50 years your *date of birth* also has to travel back to an earlier starting point. That doesn't have to be *exactly* 50 years, but it does have to be *believable* to those parts of the rest of the universe with which you interact. And that means that your mother and father need to travel back a believable amount of time to beget you. And your kids to follow similarly. And theirs. And your other antecedents. And anyone who remembers you. And their spheres of influence. And so on. It gets *complicated* for even the God(s) to resolve. So if you want to build and use a working time machine, you might want to think about making the design simpler by choosing a life of quiet anonymous celibacy!

The ultimate irony of this, and it is quite involved so I'd suggest you either think it through at leisure or just accept the conclusion, is that while time travel is perfectly possible, albeit rather tricky, as far as you or any other person or dog will be able to tell, all time travel that is going to happen has *already* happened, even if some of it might be in what we think of as the future. Therefore you personally won't be able to kill Genghis Khan because you didn't, which means that you personally won't be building a time machine (assuming that to smite Genghis Khan is what you would have used it for if you will had done it), so you needn't waste any time bothering to try, especially if you have kids.

Ah. That feels better. You probably noticed that I just correctly used the 'future pluperfect' tense, 'will had done it', which your English teacher no doubt always denied even existed. I imagine that there will be grammarians lining up willing to pay good money to be allowed to rub my tummy after that one!

Anyway, to answer your question further, while we've already established that all time travel into the past has already happened and that any paradoxes or broken house rules back or forth resolved 'immediately' (in that it essentially happens outside of what we think of as time, so certainly takes no more than 5×10^{-44} of a second), such that to us mere mortals the consequences just appear as normal everyday life, so you're never going to meet a future copy of yourself, what *would* actually be reasonably easy would be to travel into the future (relative to what you think of as the present), although once you're there you can't come back. That's just a matter of simple classical Einsteinian general relatively, in that time is relative, the consequence of which is that if you nip off to Zog at near to the speed of light, orbit and nip back, time for you will almost have stood still relative to everything back on earth, which would have continued its regular passage through time. (I would of course also note that in a universe with no house rules, your *clone* might just materialise from nothing before your very eyes and even look like you but for argument's sake ten years older, but that is *not* the same as meeting an *actual* future copy of yourself.)

This 'time dilation' effect can and has been proven to fit the model of general relativity by flying around the world with an atomic clock in a jumbo jet with an equivalent synchronised clock back on the ground: the clock on the ground 'ages differently' to the one in the air (not by *much*, billionths of a second in fact, but it *is* measurable, consistent with the theoretical mathematics, not dependent on which clock is flying, and doesn't happen when the clocks are stationary relative to each other, ie they're still whizzing through space on the earth, but neither is up in an aeroplane, although because of the contrasting but similar gravitational effects on time dilation there are other factors that influence which clock actually ages the fastest and it depends on the direction of travel!). Therefore, all you have to do to see the earth's future (and get stuck there) is *carefully* travel at close enough to the speed of light for *exactly* long enough to have the desired effect. Too fast and you might find everything goes dark because time has effectively come to an end and all the stars have gone out. Too slow and (almost) everyone you know will still be alive when you get back and will probably be somewhat miffed that several years earlier you left a note on the kitchen table to say, 'Gone to Zog. Don't wait up for me,' and then just disappeared.

So, to sum up.

You personally will not be building a time machine to go back in time because if you will had been going to do it you will already had done it previously in a few years time.

This should not in any way be confused with any inferred pre-destiny in which you wish to believe and the ensuing questions over freedom of choice and free will, which we shall of course cover more in due course and mercifully, you may be pleased to hear, far more simply.

Doctor Who (by which I mean the *real* Doctor Who) *can* travel back and forth through time as he has special dispensation from Pyror based on the understanding that he will not cause any *inappropriate* paradoxes, ie those which wouldn't resolve instantly and with a favourable outcome without ripping a conceptual hole in spacetime. Any such paradoxes present in the BBC's *fictional* series *based* on the life and times of the real Doctor Who, I was assured by a lady named 'Doomwolf of Darkness' (a friend of Mad-B who often sports a multi-coloured 'death hawk' haircut and at the time sang in a Goth band called *Mourning for Autumn*), were at the time of her comments entirely the fault of Russell T Davies, the then lead script writer and executive producer of the aforementioned fictional series. I am unable to comment with any authority on the validity of her claim and have been too busy of late to form my own opinion on the latter series since Mr

Davies decided to take a well-earned rest. I am also unable to comment on the more recent assertion of Doomwolf of Darkness that *Mourning for Autumn* are a 'big bowl of custard', my understanding being that in this context 'custard' is a hybrid vulgarity of biblical proportions based on words that I personally do not wish to use here, viz an 'initialism' derived from certain other words. How frightfully rude. Max Merrybear does NOT like bad words and does NOT intend to Google 'Jeremy Clarkson Custard' or the 'Urban Dictionary' to find out more!

Father Christmas, as we have already established, operates entirely outside of our time without ever affecting the past. Quite frankly, therefore, and with hindsight, that part of your question was arguably rather silly.

I imagine the Tooth Fairy, if he or she exists in your universe, which is of course possible, would probably either need a lot of help building a suitable flying saucer and working out the more complex astrophysical calculations, or to work outside of time in the same way that Father Christmas does. To the best of my knowledge, my universe does not have a resident Tooth Fairy, and if there's a Tooth Fairy living with Pyror outside of the Quantum String Theoriverse he or she does not appear to interface with my universe, so I cannot be sure. In your universe you might even have a small army of Tooth Fairies, some working unseen outside of time and some whizzing visibly around in flying saucers, however I would like to make it perfectly clear that the more Tooth Fairies you see whizzing around in flying saucers, the more likely you are to need professional psychological help if you wish to attempt to interface with any other people's universes. That said, you may of course be spot on in your observations; after all, it is *your* universe ... But maybe just bear the psychology thing in mind when talking to people from fairieless or low-fairy universes ...

Nerd alert!

Please note: our new understanding of time does mean that Mad-B's favourite film, The Terminator, is in fact not entirely accurate in its portrayal of time travel. That said, in $2^{\text{GOOGOLPLEX}}$ universes with no house rules, not only is it entirely feasible, but also fairly common, for a cyborg (et al) to appear out of fizzing thin air while *believing* it has travelled back to the past to change the future, when in reality it would just have materialised from nothingness in the present but sporting that set of beliefs because of the out of control nature of the universe. If that makes no sense, it may be that you're still relatively normal and you still have some work to do on your nerdity before you're welcomed with open arms at nerd conventions, so I wouldn't worry too much!

Back to normal.

Right. 'Time' for another question. Ho, ho, ho.

23) If you really are a dog, on what the heck are you going to spend any royalties or commission from sales of this book?

Woof!

OK joking aside, although if you've yet to learn to read Canine language I imagine the hilarious joke 'Woof!' probably loses something in translation ...

I suppose the first thing to mention, is that being a dog I've always tended to eschew material possessions, and while I confess that I do dislike it when writers quote obscure material in a subconscious or overt attempt to massage their own vanities, I think that at this juncture the words of Alcuin of York to Archbishop Æthelhard of Canterbury in circa 793AD are particularly appropriate: 'Let not the wealth of the world ... muzzle thy lips from righteousness.'

Anyway, despite those sage words, there are a number of things I'd like to do if I had a little human money. Here, in no particular order, they are ...

I've never really been the type for studded gold collars, although I do like to travel, albeit just by modes of transport currently readily available in my own personal universe, as I'm not at all sure humanity is quite ready yet to see a dog piloting a flying saucer. Therefore, as we shall see shortly, this basically means me travelling in the back of a Mondeo Estate, in which I have been to such far-off lands as Cornwall and the Lake District. I suppose my future travel plans may be influenced somewhat by the policy of airlines on dogs flying Business Class: as you can appreciate, I'd suffer trying to make do with the leg room in economy. I've always fancied going somewhere cold and damp that smells a lot, so I would be happy to receive your recommendations.

Mad-B had to fork out for his somewhat less than shiny second-hand Mondeo Estate when I was six months old and no longer able to fit into his previous Mondeo Hatchback, and the car now exhibits signs of wear and tear, such as holes in the seats and floor, and rust patches from neglected 'age lines' where on several different occasions in completely different car parks, to quote Mad-B, 'some stupid bugger must have built that ruddy bollard when I wasn't looking – it fudging well wasn't there yesterday,' so I might buy a newer car and let him borrow that.

There's a cottage in the Lake District unto the proprietor of which I would like to send a new bed with a letter of apology, and I'd buy Jazz a course of cognitive behavioural therapy to try to stop him defecating in the sea at Studland in Dorset, England.

Oh yes, and Mad-B's younger stepdaughter wants another pony and a Red Setter puppy called Charlie. And some stables. And everything in the Joules catalogue. And she does rub my tummy sometimes so I'll probably buy her an Oakley horse box as well, although they're 'quite expensive' so it may well have to be pre-owned.

Oh, and I need to spend £100,000 on a new roof for the church in the village of Boscombe, near Salisbury in England, and not to be confused with large and well-known suburb of Bournemouth.

I think that's it.

24) What?!

Ah. You noticed the church roof thing then. Well I *have* been to midnight mass and to a couple of special pet services, albeit the humans there present seemed to be concentrating on just a single manifestation of Pyror, which is of course fine as long as it makes them happy. Pyror likes people and good dogs to be happy. He's usually quite content to listen, however he's being addressed, even if he's called 'Mother Earth', or even something garishly functional like the 'Head of Omniversal

Services', albeit he can't always do anything about what's being discussed, at least not in this game of Quantum String Theoriverse, but then you probably already know that. And of course sometimes he doesn't like what he hears, or he discounts it because it would mean he'd have less chance of winning the Best Quantum String Theoriverse competition, but he'll listen to everyone. Eventually. Even atheists. Whether they like it or not. Of course whether that's because he's astonishingly altruistic or because he *really* wants to win the prize, whatever it may be, is a matter of philosophical debate. When I was 'dead' *he* said it's just the former but there's often at least a slight overlap in altruism versus reward, even if it's just peer kudos: it's a tricky one, that's for sure. On the other hand, would anyone have a chance of winning unless they were inherently altruistic? What do you think? By all means let me know. Terms and conditions for the submission of emails, Facebook messages or snail-mail letters as previously. I don't yet Twitter.

Oh, and talking of church, I was also *nearly* senior ring-bearer at Mad-B's wedding to Mrs Mad-B, but in the end they were slightly worried that I might drool all over the presiding Bishop (a friend of theirs who usurped the regular 'Team Rector' of the group of six parishes for the event) and also that Jazz might hump his leg and defecate in the font, so we *both* got put in kennels for ten days. Humph! The indignity!

Anyway, to answer your question more fully, the church in Boscombe is, at the time of writing, in fairly desperate need of a new roof and, being a rural church, barely has two coppers to rub together what with limited parking facilities, a small village population, and an ancient and enormous Gothic cathedral a few miles down the road. By the time the Diocesan Share has been paid (essentially to run the parishes and pay some of the clergy in the area a modest wage/stipend for what's often a seventy hour week with less cream teas than some might assume), plus various bits and bobs like insurance, there's about tuppence ha'penny a year left in savings to keep the fabric of the building intact. Lots of rural churches are in the same boat and aren't quite as awash with Nazi gold plundered from Swiss banks as certain tabloids would have us believe; at least not Church of England ones, which is the case here. Trust me: I've seen their balance sheet. In fact many rural parishes actually often run at a deficit and cling to their existence with charitable hand-outs from the diocese funded by better-off urban areas.

But back to Boscombe, and to save me having to further hone my questionable literary skills and describe the church more eloquently, here are a couple of photos of it, albeit the roof does appear far more robust and photogenic here than is actually the case if you see it in the flesh.

Boscombe is also the place where Mad-B buried his first corpse, or at least the first one that he's admitting to. He assures me that it's quite a poignant moment popping someone's ashes into the ground and covering them over, in a casket in this case, although he's done a couple by pouring the ashes in without a casket, which apparently is 'a real bitch if it's windy'. Personally, I can't think why that might be a bad thing: my mum was a bitch and everyone loved her lots and lots, especially her Little Maxi.

Anyway, the gist is that he'd dug the hole the day before (with help from Mrs Mad-B … in fact she did most of it, and certainly the bits that involved bending down on all fours and scraping out the earth and rocks with her bare hands), casting the larger stones into the shrubbery around one of the more modest trees to the left of the church, under the bows of which a lady we'll call 'Mrs X' to let her rest in anonymous peace was to be lain to rest next to her husband who had departed this world some years earlier, while retaining the best of the soil in some discreetly-stored buckets for use during the interment.

During the course of the digging, what with it being slightly under the outer branches of the tree, they'd encountered a few relatively minor roots, up to an inch or so in diameter, one of which actually proved to be the fibula (calf bone) of someone buried there around four hundred years ago (as estimated by Mrs Mad-B based on the colouration and state of the bone: I defer to her knowledge of the subject of the timescale since the original burial, as I've always tended to dig bones back up rather more quickly myself!). These bone encounters are a not uncommon event when digging in older cemeteries where people used to be buried in unmarked graves, although it was not until Mad-B's third hole, while chivalrously actually doing some digging himself, that he encountered a skull, or at least a skull cap; a sad event as the subject of that burial was also supposed to be buried with a loved one's ashes (in this case in a double-depth grave, whereas Mr and Mrs X were side by side) at a site previously unmarked by any gravestone.

Unfortunately it seemed likely that the wrong place had been identified by the funeral director (ie the skull cap would probably not have been found if the location were correct, as insufficient heave of the ground would have occurred to bring the skull cap into that location, as it had previously been dug just a few years earlier. That said, the skull cap *may* earlier have been reburied in situ on top of the ashes if the gravedigger were ignoring protocol or in something of a rush for whatever reason, so if you're the ghost of either party please feel free to rest in peace rather than wandering the earth looking for your loved one's remains: we had to go ahead with the interment to avoid leaving the bereaved more battered and bruised than they already were.

Anyway, with the calf bone of the gentleman, now nicknamed Erasmus, reburied several inches deeper but in the same place as is the accepted procedure, Mrs X's plot was finally shipshape with the flat gravestone now arrived to be placed on the grave once the casket had been interred and the grave backfilled to the requisite level to nestle the stone flush with the surrounding turf. No worries. Next day there would be a few prayers at the graveside, then the mourners would lay the casket in the grave before filing into the church for a simple service of thanksgiving for the life of Mrs X, the service scheduled to last for around 25 minutes.

As soon as the church door closed for the service, Mad-B would leap unseen from his (t)rusty Mondeo, parked earlier at a suitable if not entirely subtle vantage point, gather his tools and watering can, fill in the hole with the discreetly-stored soil, position the gravestone, neatly pack the finer soil around its edges, pat down some turf at the edges to complete the aesthetics, then wash it down if required as many gravestones will become slightly muddy during the operation. Mad-B would then scurry back to his car and be off in around eighteen minutes, ie with seven or so minutes to spare before the service was over and the mourners filed back out to pay their last respects at the immaculate graveside before heading off far and wide around the country.

Except that with all the bones, stones and branches that had been removed there wasn't even remotely enough good-quality soil left. This was no time for messing around and leaving the mourners traumatised by a sunken gravestone, so quick as a flash, thinking on his feet for once, Mad-B hastily dug a few shovels of topsoil from the far side of the tree, topped Mrs X up with the good stuff, then scrambled in the bushes and under the tree gathering up discarded stones and moss to back-fill the impromptu hole: it would have been far too cruel an irony if someone had fallen into it and fractured their fibula!

By the way, Erasmus. Mad-B says he sincerely hopes that your soul isn't 'hopping' mad. Puns are *not* Mad-B's forte, although I fully accept that I am in no position to criticise.

Anyway, after just 23 minutes, Mad-B jumped into the car, his task duly and diligently completed, and drove off.

There is a further irony in this tale that Boscombe church is where Richard Hooker served as Rector and wrote much of his 1590s work *Of the Laws of Ecclesiastical Polity*, one of the most influential works from one of the most influential characters from the early Church of England: essentially a treatise on Church-state relations written at a very formative time for Anglicanism, after the argy-bargy with the split from Rome that Henry VIII's shenanigans had caused, but it also dealt comprehensively with issues of ethics and tolerance. Throughout the work, he made it clear that philosophical theology should be concerned with ultimate issues, rather than petty quarrelling and injustice.

It is generally believed that Richard Hooker was buried in Bishopsbourne in Kent, where he later served as Rector until his death in 1600. But of course we'll never know for sure that his physical corpse didn't return to Boscombe for a more covert interment. It's possible, and far too late to prove otherwise, that Erasmus was actually Richard Hooker in disguise. Boscombe is in the Bourne Valley. Two Bishops live there (at the time of writing). *Bishops ... Bourne.* At the *very* least it's possible that, emulating the practices of the earlier church to share 'relics' as 'tourist' attractions, such as the bones of famous or saintly people, Richard Hooker's fibula was buried at Boscombe. It'll drive the conspiracy theorists nuts if nothing else, and Mad-B has just fixed the toilet in the church, so it's now ready for a Da Vinci Code style influx of pilgrims and tourists!

And then a few months later, I 'died'. And then, through his floods of tears, I got Mad-B to make a deal with Pyror (about whom I first spoke to Mad-B as I lay fouling myself on the kitchen floor) that if I lived until 10 years old he would type this manuscript for me and furnish Boscombe with a new roof.

I think Mrs Mad-B will be a bit surprised when she reads this, as he hasn't mentioned it to her yet. Let's hope she's 'quite' understanding!

25) I just saw an episode of Horizon on BBC2 entitled *Is Everything We Know About the Universe Wrong?* They talked about The Standard Model of Cosmology, Inflation, Dark Matter and Energy, and Dark Flow, and showed me exactly how the universe was created with commentary from some of the finest Cosmologists and Theoretical Physicists on the planet. What have you got to say about THAT Mr Merrybear?

Please, if you would be so kind, even if maybe you skipped or skimmed over a few things before, read this bit and read it carefully, because this is what science, and rest assured I'm generally a big fan of science, tells us is definitely happening, although not why, and if you want to take away one single thing as the weirdest thing ever known, this is it: 'Trillions of stars in thousands of galaxies spanning a billion light years of space are travelling in a supposedly impossible way towards something *outside of the universe*.'

Please read that bit after the colon again. It's definitely worthwhile forming a picture in your mind.

This, in my opinion, is evidential proof beyond any reasonable doubt that there are one or more things that are *outside of our universe* and which *directly and massively* affect what *happens to us*, and probably things we may find scary because, unless something other than simple pan-omniversal gravity is causing the attraction, when combined the things would have to be many trillions of times the size of a fairly large black hole, and if it's *not* pan-omniversal gravity it's potentially even more scary!

There are a couple of other sinister goings-on as well, both equally irrational. One, stars *cannot* rotate around the centres of their galaxies in the way that they *do*. Those at the outside travel at much the same speed as those closer to the middle, which by any normal maths or physics should mean that the galaxy flies apart: this can NOT even remotely be explained by black holes, although doubtlessly black holes help to provide something hefty around which to rotate, it's just that black holes don't in any way explain the weird speed thing! Two, for the last seven or so billion years the *rate* at which the universe is expanding has been increasing. It's not just still expanding; it's expanding faster and faster, which is the exact opposite of what happens in a regular big bang.

The universe is *impossible*. Impossible in at least three completely different ways.

Until now.

Allow me to elucidate, although if it's all too scary already you can skip the explanation and proceed to the next question. It'd be great if you can hang on in there though.

I happened to be warming my weary bones by the fire after a particularly fine dinner of duck in a rich port and blackberry reduction, and was fortunate enough to catch that episode of Horizon myself. I have two main observations, other than the fact that I do indeed laud much of the thinking and intricate mathematics behind it all, which I shall of course expand upon, including a little light background information where needed for any readers unfortunate enough to have been otherwise engaged during the screening.

Observation one: their astute interpretation, theories and speculation, including the radical new concept of Dark Flow, *reinforce* the teachings of Dog, and enhance my assertion that science and religion can live in perfect symbiotic harmony and, observation two, despite the way it was phrased, they didn't actually tell us *how* the universe was created, or indeed *why* (although explaining *why* was not one of their claims), unlike in *The Word of Dog* wherein we learn full and unabridged answers to both (insofar as such answers have any meaning within the omniverse – there will almost certainly always be theological debate about the exact nature of the outside of the omniverse, although there is at least enough information in this book for you to be able to fine tune any grey areas in your beliefs or meta-beliefs). Rather, what the episode of Horizon actually told us, in all probability correctly from the point of view of people inhabiting one of the more popular science-based universes, although perhaps not quite so much from the perspective of the more radical fundamentalist Creationist types of universes, is how the universe in question is likely to have behaved *once* it had been created. And indeed 'after it happened' is what the Large Hadron Collider seeks to help clarify, although Dark Flow is still *highly* likely to remain a beautifully terrifying mystery to anyone not reading *The Word of Dog*.

Let's précis, explain and discuss their view in terms that even Jazz could understand (by which I mean the view where the universe was created 13.7 billion years ago in a 'bang' – this discussion does not preclude you believing otherwise if you so choose, nor indeed being correct in your belief), and not only that, but we'll get you, the mighty Pingpongor, to demonstrate Dark Flow in action! ... Unless you, Pingpongor, fancy a nap or a break, in which case feel free to skip to the next question/discussion point, secure in the knowledge that there is nothing about even the most avant-garde physics/maths/astrophysics that in any way conflicts with the information bestowed upon me by Pyror to share with you should you be interested, which I hope you are.

So, the original Creationist model of the universe started along the lines of Genesis 1, which reads quaintly from the King James version of *The Bible* but quite frankly is a nuisance to quote in books in the UK because Her Majesty still owns the copyright, albeit you can Google it on trade websites based in other countries (eg www.biblegateway.com) or download it, but basically it started with God creating the heaven and earth, giving it form, creating light, dark, day, night, sea, land, plants, fish, birds and beasties etc. Oh, and man. And woman.

Incidentally, something very similar to that may have been read by 'Erasmus', albeit due to the particular timings it would probably have been from perhaps the 'Bishop's Bible' which (along with various contemporary alternatives) was a post King Henry VIII predecessor of the much-revered King James version but I digress.

Anyway, there's no reason why that might not be 'true', even if an average non-Creationist might interpret the wording as metaphorical rather than literal. Many people: many interpretations. From true, verbatim et literatim, to complete gibberish. It's up to you. We're just thinking here of how things were originally presented.

Next, some people decided that they wanted a slightly

different slant on things, so along came, 'Hey, it looks like there might have been a ruddy great big bang. I have no idea why though. It probably just happened for no reason. I think I'll believe that.'

Now we have *The Word of Dog*. Although a full and unabridged definition of what that means will have to wait a little while, as we've got a few more important building blocks to consider, however just to solicit your continued interest I can at least assure you that the definition will come from a source you may not be expecting ... OK, it's you: you will give the definition ... So there's no need to flick forward to see what's coming (in a later chapter) ... No there isn't a quiz, so relax ... OK, to save you having to think too hard I'll just tell you the definition and you can decide whether to agree or not and take the credit. Tough audience; but at least Max Merrybear is an 'all about sharing the love' kind of dog!

The gist of the current science-based observations, where 'we' refers to Mr and Mrs John Scientist and their chums and/or anyone who wishes to associate themselves with that particular clique, goes like this: 13.7 billion years ago there was a biggish bang, but not quite as big as we thought in the 1980s, then after a while, about 10^{-34} of a second or so (about two billion time units), the universe expanded by a quadrillion quadrillion times really quickly (much faster than the speed of light) for some reason, so in effect the big bang happened just after the biggish bang, almost like the biggish bang being a detonator that set of a bigger bang, a bit like blue touch paper in a firework like, for example, a rocket; then after 380,000 years or so atoms started appearing out of the fizzing energy which we can 'see' in the Cosmic Microwave Background (as long as we've got the maths right), then after a billion or so years gravity started and first-generation stars started to form (ours being a second-generation one), so apart from a few impossible things like galaxies not flying apart, the universe speeding up seven billion years ago, and that really weird and scary thing with the royal garden party going on in another universe to which trillions of stars have been summonsed, we think we know roughly how the universe has been acting since it was created for some completely and utterly even more inexplicable reason or other.

That is why any *sane* scientist is likely to embrace *The Word of Dog*.

Let's play a game: a new round of nothing less than Ping-Pongoverse, starring you, the mighty Pingpongor!

Indulge, if you will, my suggestions and:

- Decide to play Ping-Pongoverse and get a small box, say a jewellery box.
- Decide the box is too small and get a bigger one, say a shoe box.
- Toss some coins and deploy some ping pong balls.
- Keep tossing and marvel at the way your house rules gently cajole some of your ping pong balls into moving together and forming clusters of ping pong balls.
- Gasp with glee as some of the clusters start spinning, and marvel as your kitchen light seems to bounce off them.
- Notice the box is getting full and decide to get an even bigger box.
- Notice the new box is getting full and decide to get bigger and bigger boxes whenever you need to or even if you just

happen to feel like it.

- Decide to thin the box out a bit by henceforth taking out any ping pong balls that land on square 13 and putting them in an orderly line on the kitchen table, perhaps so that you've got some spare ping pong balls so that you don't run out if you keep deploying new ones inside the box.
- Decide to thin the box out a bit more by insisting that any ping pong ball within two squares of square 13 in any direction is not allowed to move further away on any given turn, just stay the same or get nearer.

Now let's merge the two:

- Decide to play Ping-Pongoverse and get a small box, say a jewellery box (there's a small bang).
- Decide the box is too small and get a bigger one, say a shoe box (expand the universe a quadrillion quadrillion times really quickly).
- Toss some coins and deploy some ping pong balls (start some energy fizzing).
- Keep tossing and marvel at the way your house rules gently cajole some of your ping pong balls into moving together and forming clusters of ping pong balls (after 380,000 years matter starts forming from the fizz).
- Gasp with glee as some of the clusters start spinning, and marvel as your kitchen light seems to bounce off them (after about a billion years some first generation stars form and give off light, possibly bestowed upon them, whether conceptually or actually, from 'outside the box').
- Notice the box is getting full and decide to get an even

bigger box (after seven billion years or so, start doing some even quicker expanding, there being some currently 'emerging' evidence that occasionally 'nothing' can split at a quantum level into a positive and negative energy 'parcel' for just a few units of time which overall still equates to 'nothing' but when they recombine they don't always cancel each other out, but rather the negative becomes positive and so energy/matter can be created from 'nothing' as deferred here from Chapters Three and Seven, and that this creation of energy/matter is involved in the mechanism for the increase in the rate of the expansion of the universe; that said, if you ask me it might well be that the negative 'flows' into a different universe leaving the positive here as new energy/matter thus emulating the non-destructive recombining but with a balancing anti-universe; and yes, it is all a bit weird, but then the universe does appear to have come from *somewhere*!).

- Notice the new box is getting full and decide to get bigger and bigger boxes whenever you need to or even if you just happen to feel like it (carry on expanding the universe willy-nilly).

- Decide to thin the box out a bit by henceforth taking out any ping pong balls that land on square 13 and putting them in an orderly line on the kitchen table, perhaps so that you've got some spare ping pong balls so that you don't run out if you keep deploying new ones inside the box (dark flow begins, perhaps though not necessarily for reasons 'outside the box').

- Decide to thin the box out a bit more by insisting that any ping pong ball within two squares of square 13 in any direction is not allowed to move further away on any given turn, just stay the same or get nearer (dark flow becomes obvious enough for an educated ape-descendant to notice).

Sometime later, you may also have seen Horizon's follow-on programme: *What Happened Before the Big Bang?* This, in my canine opinion, was in several ways a somewhat less well constructed programme than the previous episode, although largely reinforcing some of the ideas, such as inflation, when it wasn't also belittling them with questionable or discredited opposing theories. There were however, as I shall discuss in a moment, a couple of *very* important pieces of information hidden within the confusion.

First, however, here are some quotes from the programme, all from respected scientists:

- 'Time did not exist before the beginning. Somehow time sprang into existence ... which may be a logical contradiction.'
- 'There is suddenly a big bang. That is impossible. I don't believe that at all.'
- 'It gives us a sense of origin ... It's a comfort to know exactly where we came from.'
- 'Ten years ago, "before the big bang" made no sense, but today the certainty has gone.'
- 'It's a problem. It's all effect and no cause.'
- 'It's preposterous. How can it be that everything comes from (absolutely) nothing?'
- 'The Big Bang itself is a flawed concept but one that holds tantalising clues to the *real* story of creation.'
- 'The universe appeared out of the cheese of ... eternal

inflation.' (Yes, that was an actual quote!)
- 'In mathematics, invoking infinity is the same as giving up ... something has gone terribly wrong.' (I firmly agree!)
- 'The Big Bang was not a bang at all, it was, rather, a Big Bounce. It's a surprising thing, a bouncing universe.'
- Dr Laura Mersini-Houghton: '(String theory) seems to provide an elegant solution as to why our universe emerged in the first place: you do end up with a high energy big bang. It's (almost) too simple.' (Then narrator's commentary) 'On the face of it, the theory looks much like the others; it predicts a multiverse and at least one Big Bang but it stands out in one crucial respect: it doesn't commit the scientific sin of assuming initial conditions ... it assumes nothing at all ... the other remarkable thing about the theory is that it fits with three observations; phenomena which have defied conventional explanation: there's an unexplained patch of nothing, the so-called "void" in the Cosmic Microwave Background; great swathes of galaxies have been found to be moving in the wrong direction (ie Dark Flow); and there's something "odd" about the temperature in outer space ... According to Mersini-Houghton all of these effects are explained (in precise detail) by the presence of neighbouring universes.'

Alas, what the program *didn't* do, using any of the divergent and conflicting theories presented, is explain where the universe(s) came from. Luckily, *The Word of Dog* bridges that gap. Here is another part of the bridge.

If the universe *is* bouncy, which basically means it either gravitationally implodes then re-explodes, or alternatively at some stage in the distant future possibly, and this is *very* maths-nerdy and liable to cause the odd argument but is entirely possible if the omniverse does indeed prove to be a pan-dimensional hyperbolic Möbius ouroboros, it expands so far into hyperbolic four-space that it loops back round on itself and re-emerges from where it began (which is impossible based on what we *see*, but then what we see with our eyes doesn't cover what goes on in hyperbolic spacetime, although conceptually it's a bit like walking around the world with what you see looking essentially flat, but one day arriving back exactly where you started, only with more dimensions and more weirdness). This raises an interesting discussion point: if the universe bounces, the energy strings would be rearranged for sure, but in essence they might well still be the *same* energy strings, so theoretically that would provide a mechanism for 'remembering' things that happened in 'previous' universes using different house rules, so some of the $2^{\text{GOOGOLPLEX}}$ possible universes that some people seem to think they inhabit, may technically be because once upon a time that's exactly what *did* happen. So for argument's sake, maybe in a *previous* version of the universe there was a Jesus called Mithras, rather than it being true for a concurrent version of *this* universe.

Muse on that at your leisure if you will. It's one of the most important things you will ever read anywhere, in that it provides a flawless link between science and all religions.

And that's why everything we know about the universe is now right. It was just a simple matter of working out what the house rules might be and then deciding that the universe may be bouncy: rather like knowing what the rules of the game Ping Pong are and how the balls are likely to bounce while you're playing.

26)   Intriguing. So if we assume for now that the double slit experiment, which I, the mighty Pingpongor could, if I so chose and I had the right equipment, do on my own kitchen table, demonstrates that, however stupid or impossible it sounds, an electron can be in two different places at the same time, and that this gives credence to the assertion, which actually has real and demonstrable applications for the cutting edges of quantum computers, again however impossibly stupid it sounds, that a quantum bit (Qubit) of information actually can be a zero or a one *at the same time*. Therefore with two binary Qubits they can have the total value of zero, one or two *at the same time*, not unlike when a lady both wants and does not want a large bar of luxurious chocolate at the same time, so it is feasible that the whole of quantum mechanics is actually a little jokette of Pyror's, in that similarly Pyror himself may be one god (ie Pyror), more than one god (Pyror and chums, who some people may think of as a trinity with saints, or perhaps some of the more collective Norse, Roman or Indian gods for example), or zero gods *simultaneously* (say if he'd popped to the shops for Mrs Pyror and so was not actively engaged in the Quantum String Theoriverse at that particular 'time'). Therefore science and all religions including atheism can be concurrently correct, so it's actually a *bit silly* stressing and fighting about who's right. What have you got to say about THAT Mr Merrybear?

That would seem to be a reasonable deduction, given that it concurs entirely with where I've been hoping to gently guide you. Congratulations. You've certainly improved the clarity of your thought since that incident with the flying saucer!

Nerd alert!

The only minor observation I have is that your proposed model makes no reference to quantum superposition, which technically provides a mechanism for Pyror additionally being both 'real' and 'imaginary' at the same time, although that may add a level of complexity which some readers may find distressing, so I can see why you'd want to avoid too much detail, but as long as you're aware of the concept that's fine, and of course in that sense we are using the word 'imaginary' in its mathematical context, which loosely speaking defines a number which is 'impossible' but also essential to the everyday use of electricity, ie the square root of a negative number in whichever dimensional vector you happen to be interested.

Back to normal.

So. Yes. Pyror is none, one and many. At the same time.

27) How many intelligent civilisations are wiped out every day, and why does Pyror let that happen?

Let's dabble with a few numbers based on what most astronomers would have us believe, and in this case the numbers appear entirely reasonable within the context of the universe most of us seem to inhabit. So, and this may just freak you out COMPLETELY, one very real possibility is that there are 5.0184516×10^{22} stars in the universe (spread among some 100 billion or so galaxies for the sake of having at least one round number!). Why would that freak you out? Because that's the number of Carbon 12 (ie normal carbon) atoms in EXACTLY one gram of normal carbon, ie the carbon that behaves weirdly in a number of unique ways and means that life as we know it is possible.

I'd ponder that for a while longer if I were you: there are quite possibly *exactly* the SAME number of STARS in the universe as there are ATOMS in one gram of the thing that makes organic life possible. TO THE VERY ATOM! Odd? I think so. It could just be a coincidence of course. Quite frankly, it's more than just a little bit weird though. Almost as if someone's trying to tell you something. Still, don't take my word for it; I am just a dog, so by all means draw your own conclusions!

And of course everyone and their dog seems to be discovering a new planet around some far off star or another these days, so let's assume for the sake of having exceptionally simple maths that for every 2,000 stars there is just one planet where Pyror's house rules have permitted life to flourish in some form or another (and from the rather splendid movie *Jurassic Park*, we do know that 'life finds a way' so it could easily be a *lot* more, or indeed less!). Let us also assume that most of that life is really dumb, say nothing more intelligent than a washing machine powder advertising executive, and so that for every planet with life on it, only one in every 2,500 has 'intelligent' life, although please don't make the common enough mistake of taking the definition of intelligence to mean 'as intelligent as me', which is of course a profoundly stupid way of defining it as we shall soon see. I would also like to apologise for singling out washing machine powder advertising executives for derision. I very much dislike belittling anyone purely because they live in a different universe to normal people, but Mad-B insisted. I think he has issues: after all, in some universes it's just possible that grass stains, grease, curry stains and blood *will* disappear in a cold wash ...

Let us also assume that every 300 million years an asteroid zaps the average planet, or a nearby star goes supernova, or a Large Hadron Collider goes berserk and kills everyone, and only intellectually simple life forms (if any) remain to start again almost from scratch. Rather conveniently there are about 10,000 trillion seconds (10^{16}) in 300 million years. Ergo, the answer to part one of your question is: 'The total number of stars, divided by the number you get when you multiply the habitable planet distribution by the intelligent life form distribution by the frequency of zapping.'

Therefore, if we do the sums, using a spot of entirely appropriate and modest rounding for simplicity:

$$\frac{5 \times 10^{22}}{(2 \times 10^3) \times (2.5 \times 10^3) \times 10^{16}} = \frac{5 \times 10^{22}}{5 \times 10^{22}} = 1$$

Therefore one intelligent civilisation is wiped out *every* second, or 86,400 per day. If you wish to multiply that by however many *concurrent* universes you've decided to believe in, by all means do so. Here we're just considering the universe that appears most popular amongst readers of *The Word of Dog*.

And that's *civilisations*, not individual *life forms*.

Using my knowledge of *Star Trek: The Original Series* (ie in Mad-B's opinion just 'Star Trek', the bit after the colon being an inconvenient addition caused by later improper and incorrect series) to extrapolate this data, where all intelligent life in the universe comprises bipeds (bite me!) on Class M planets, inferring that the average Class M planet has a current population (unless it's just been wiped out by a phage) of around 3.5 billion (although obviously Star Trek takes place hundreds of years in the future when computers will have become huge again and once again started going 'ping' when a light flashes, and coincidentally populations will have stabilised to that similar to earth in circa 1966 to 1969 when the original series was written), and discounting any life forms so stupid that they allow themselves to be beamed down onto polystyrene alien planet surfaces during their first day on the job and wearing an immaculately laundered grease-free red tunic, we can reasonably say that this equates to exactly 300 trillion intelligent life forms wiped out every day. By inference we've already shown that this is the same as everyone on earth in the late 1960s being killed every second, excluding any new employees wearing a red tunic (cf Montgomery Scot, Chief Engineer, aka 'Scotty', who was wont to wear a red tunic but was a seasoned and valued employee), not that they wouldn't be killed, it's just that they wouldn't be counted as intelligent life forms given that by inference they would previously have thought it was a good idea to beam down onto an unknown alien planet surface having never previously met an alien and while wearing a bright red tunic emblazoned with the motto 'Dumb-Ass Bug-Eyed Monsters Can't Shoot Straight!'

Obviously the actual *exact* number of intelligent life forms wiped out will vary based on your own personal definition of intelligence. I'd *like* to think that after reading this book you'd include at least some canines within your definition. Maybe some dolphins, cows, lesser primates and maybe guinea pigs as well, given that we mentioned in Chapter Eight that guinea pigs are surprisingly eloquent little rodents? It's really up to you how you define it, although naturally the more life you think of as intelligent the more is being wiped out every single second, so that could very easily bump the figure up to a quadrillion a day (a number, equal to 1,000 trillion, that rather spookily we're about to use again for something related!).

Similarly, you might judge intelligence more harshly, in which case a smaller amount of intelligent life is being wiped out. You may even decide that everyone and everything less intelligent than you counts as stupid, which of course by definition means that everyone else you count as intelligent is actually no less clever than you, which is in fact decidedly ironic, because logically, if that be your stance, by your own system of measurement you would be the equally least intelligent life form in this or any other universe anywhere, so my advice would be not to judge others too harshly, lest by association and logic you actually be judging yourself!

Oh, and by the way, when I say 'nearby supernova' I mean any supernovae near enough to wipe out intelligent life on a Class M planet. They chuck out a LOT of nasty radiation, although this varies massively depending on the type of star that's exploding and in which directions the main radiation bursts travel, as by no means does it occur evenly. For example, the star known as IK Pegasi (HR 8210) is about 150 light years away; about 1,000 trillion miles. That sounds like a long way, which by our usual earthly standards it is, and only a truly miniscule amount of the radiation generated would actually reach us; but if it's vaguely 'pointing' towards us when it goes bang, which it might well do somewhen in the next hundred million years, not that I'm saying it'll be next week or next year, let alone two days before the day after tomorrow, we all die. Even people in Australia (or, for dramatic effect for my Antipodean readers, even people in Europe). Everyone. Gone. Fried (conceptually – it's much more likely to be due to a reaction that messes up the atmosphere rather than an actual instantaneous frying pan scenario). The odd cockroach might survive, but it'd be curtains for mammals.

Why does Pyror let this happen?

Well maybe, just maybe, you can't have a Quantum String Theoriverse with life that doesn't occasionally also have random death and destruction without it being a really *boring* Quantum String Theoriverse, where the complete lifecycle and position of every piece of energy or matter is defined for every unit of time of its existence according to a set of ridiculously burdensome and cumbersome house rules that totally remove the freedom of choice and free will of anything contained therein that might otherwise exhibit signs of life. A gilded cage is still a cage, and living in a cage sounds *bad*. It's a bit like the difference between going to a firework display with all the fun of the bangs, and the random fizzing when something flies off in a way that it isn't meant to, where there's always going to be a slight risk that you're going to get burnt, and the safe but boring option of looking at a photo of someone else enjoying fireworks. Maybe in the best Quantum String Theoriverses intelligent life just has to take its chances and trust that Pyror will sort it out properly *afterwards*. Otherwise, could you really call it life? I think not, but ultimately that's for you to decide … and maybe to email me your thoughts …

28) I think that the earth is flat and at the centre of the universe. The sun and moon orbit around it. There are no planets other than our flat earth. The stars are pinholes in the firmament into heaven and are several miles away, not trillions of miles or light years, just a few miles, and through those pinholes the gods can see our every deed. What have you got to say about THAT Mr Merrybear?

Hey, dude, it's your universe, and that model worked just fine for a large part of the Roman Empire (including Lucius Vorenus). However please be careful that you don't fall off the edge of the world if you decide to take a round-the-world trip; the keyword there being *round*, ie the shape, within certain loose definitions of the word and avoiding too much brain overload by considering that space itself is curved, that most people seem to think the world is today.

Of course there's nothing to say that Pyror might not have created a set of house rules that allow you to inhabit what I shall euphemistically call a more 'exotic' universe than most of the rest of us, and to interface with ours through what in simplistic nerd-speak would probably be called a worm hole between the two universes that are parallel to the extent that they inhabit the same time, although they are otherwise largely dissimilar.

So. To summarise. It sounds like you may be from a completely different universe altogether, *but that's OK*.

29) Is Pi (π) wrong?

Oh goodness me yes, although not in a bad enough way to make much difference, except for anyone calculating or worse still *learning* it to too large a number of decimal places. Those individuals (and arguably most people concerned with the branch of mathematics known as Fractals for that matter) are, as we shall see in more detail very soon, wasting their time, which is the sole and exclusive fault of some mathematicians' obsession with infinity which, as we've already mentioned, is 'the same as giving up'.

Let me make this *perfectly* clear: anything relating to this omniverse with 'infinity' in the answer is WRONG! ... Unless perhaps the question was: 'What's the most stupid word there is?'

Mild ongoing nerdisms for a while!

Consider the universe, or rather our standard everyday three-dimensional expanding balloon model of it, because that'll do nicely here, so we really don't need to worry *too* much about the fact that as mentioned earlier the rate of expansion is increasing, so conceivably the speed of light may also vary over time, although that may well be tricky to observe or prove and indeed I am not aware of any data to suggest that it's happening at the moment.

As you may remember from the 'full scale nerd alert' just over three quarters of the way through Chapter Two, after 10^{200} years, where a unit of space is roughly 1.6×10^{-35} metres, the speed of light is roughly 300,000,000 metres per second, and there are roughly 31.5 million seconds in a year (we used 33⅓ million to keep the maths 'simple'), and given that the formula for the volume of a sphere is $4/3\pi r^3$, the most units of space the universe will ever contain at any given moment in three dimensions is very roughly 10^{755}, which we somewhat arbitrarily multiplied up to 10^{759} for any special nerds demanding a hyperbolic spatial dimension.

Therefore, and while I defer to an appropriate nerd for an exact number, we can absolutely definitely state that anyone bothering to calculate or learn π beyond, for argument's sake, 760 decimal places, is actually getting the answer wrong because it can no longer possibly have any practical application or valid meaning even in $2^{\text{GOOGOLPLEX}}$ possible universes (given that energy strings, if you decide to believe in them, inhabit inaccessible lower level internal dimensions in much the same but different ways that molecules in the air inside a ping pong ball do, so in the internal dimensions π loses its meaning and effectively becomes a spherical peg in a linear hole).

So, by all means calculate and learn π to 760 decimal places if you like. But then get on with life and enjoy yourself. And in terms of the most popular current three-dimensional universe, 40 decimal places is more than enough to calculate the circumference of the universe to within a degree of accuracy somewhat smaller than one hydrogen atom, so I've never bothered to memorise more than 40 places myself. Here's π to 760 decimal places just in case you ever need it. I do hope that helps. I'll shrink the font a bit, because quite frankly it's astonishingly dull and very easy to Google.

3.1415926535897932384626433832795028841971693993751058209749445923078164062862089986280348253421170679821480865132823066470938446095505822317253594081284811174502841027019385211055596446229489549303819644288109756659334461284756482337867831652712019091456485669234603486104543266482133936072602491412737245870066063155881748815209209628292540917153643678925903600113305305488204665213841469519415116094330527036575959195309218611738193261179310511854807446237996274956735188575272489122793818301194912983367336244065664308602139494639522473719070217986094370277053921717629317675238467481846766940513200056812714526356082778577134275778960917363717872146844090122495343014654985837105079227968925892354201995611212902196086403441815981362977477130996051870721113

Back to normal.

251

So, to summarise, infinity is as stupid as π. Say that out loud: I have a feeling it may constitute some sort of joke in certain human languages ... Oh wait ... that's 'as easy as pie' ... I confess that 'infinity is as easy as π' wouldn't make much sense at all ... Is it too late to persuade you that it sounded funny in Canine ...? I confess that translated humour is *not* my strong point ...

30)   Satan: is he all bad?

Satan wants to violate the Tooth Fairy, kill Father Christmas, eat Rudolf and knock your ping pong balls all over the floor. That's *quite* bad. I think the main problem, and I'm just being theological here rather than saying it's the one and only truth, is that conceptually he's like the proverbial drunken uncle at a wedding reception who's completely bladdered on cheap cider and wants to destroy Pyror's Quantum String Theoriverse because he's angry at his own failure to come up with any house rules of his own that might stand a chance of winning the competition.

Whether that makes him *all* bad is a matter for personal reflection, although I submit that it would be a strange person indeed who *really* wanted to worship a drunken uncle at a wedding reception, even if they do like sinking a few pints themselves now and again and have no particular righteous objection or pious indignation when one of their chums puts *Scars of the Crucifix* by Deicide on the internet jukebox of the local public house.

At least that's how I see it.

What do you think? Emails as usual if you please ...

Chapter Eleven

MAQs III

Hey, that's my name: MAQs III ... Maqs III ... Max-iii ... Max-eee ... Get it? BEAR. Oops, blah, blah, blah.

So anyway, in this chapter I'll be proposing answers for your consideration to More Asked Questions III, ie another bunch of questions et cetera et cetera. And these are all BIG ones, at least in terms of importance. Apart from perhaps the first one, which is nevertheless important in the sense of us getting to know each other a bit better, thereby cementing our lifelong mutual bond of trust. And we've already had quite a few BIG ones. Size-wise this group averages out much the same as the other groups of important questions, give or take about 20% based on an approximate 'average' word count of circa 10,725 words. Funny that. Almost like it was a spot of luck. Or divine intervention. Whichever you prefer really. You might even think I planned it that way just to keep my publisher happy, although that's almost too absurd to contemplate. And they aren't exactly the same, so there's just enough ammunition to keep the conspiracy enthusiasts happy.

And of course a true nerd would now be chuckling to him or her self having thought, 'Ha ha. In binary, decimal 10,725 would be the palindromic number 10100111100101.'

It's not that funny for a normal person though. That said, it'd be quite a good Qubit jape to play if you were a deity and therefore able to influence such things, although as with the number of stars and carbon atoms in Question 27 (which is *far* more dramatic!) obviously only you the reader can decide if that's what happened here …

31) What's the longest postcard you've ever written? Give your answer in terms of the number of words, rather than the actual physical length of the postcard.

That's an unusual question, I have to say, although by no means are you the first person to ask me that. I'm sure that should tell me something more about human nature, although I confess that currently I'm baffled. If you have any ideas, answers on a postcard please … by which I'm sure you realise I mean email maxmerrybear@gmail.com; Facebook messages are auto-forwarded to my Gmail, so by all means select your preferred communication medium.

Anyway, I trust you'll excuse the literary standard of the work in question; I had only just learnt to write at the time. To set the scene, just over three years ago, shortly after my recovery from testicular cancer and just before he met Mrs Mad-B, Mad-B took me for a few days to a hotel named Dalswinton near St Mawgan in Cornwall, where I was allowed to sleep in his room but, for some unknown reason, not allowed to eat in the restaurant. During this break Mad-B took *many* photos of me, with the best two now on the front and back covers of this book (unless the cover design got changed by the publisher at the last minute, in which case you may need to go to Facebook). At the time, I was also somewhat 'unusual' for a Newfoundland, in that prior to Cornwall I *hated* the sea; and I really do mean *hated* – wild horses couldn't have dragged me into it. The only other major point of background interest not covered in the postcard was that on the day we left, Mad-B kissed the (somewhat older) chambermaid on the cheek. We subsequently found out that the next day she fell down the stairs and broke both of her ankles. He sometimes has that effect on people.

Anyway, here's the emailed 'postcard', which I entitled 'Postcard From Max'. It really is funny to look back and see how immature my writing was in those days. Perhaps you look back on things you wrote when you were only 6½ years old and find them funny too. Still, it was at least a valuable exercise in that it certainly made me realise I'd need some practical help with any future ventures that needed a lot of typing and editing!

'Hello everyone, it's Max. My daddy just took me to a village in Cornwall near Newquay for a few days and I had the bestest time, so I thought I'd pinch his email and tell you about it, plus send a few photos of me that he took. The hotel (eight rooms and its own field) lets dogs stay for free in their humans' rooms, and my daddy said he'd be happy to send more details and photos for anyone who might be interested. The food was great and everyone and their dogs were friendly. Not that I ate MY food at all, but more of that in a minute.

'For the first two nights, there were four male humans staying, so I assume they were on a stag weekend. They must have been poor though, because they had to share two double rooms and didn't drink much. Still, they seemed to be having a gay old time and they liked rubbing my tummy, plus they were quite chatty. I think my daddy fancied a cigarette at one point, but for some reason didn't ask the one who smoked if he could bum a fag.

'After dinner each night, the hotel man or lady came round with bic-bics, so I put on my very best "I haven't had one yet" face. It was ever so funny, and I certainly got lots more than some of the little dogs. Also, after three days of not eating my dog food, my daddy took pity on me and bought me chicken nuggets and chips for lunch. After that I gave him my best, "Oh daddy, please tell me the food isn't over look," so he took me for a cream tea in the afternoon and I had two huge scones, one with cream and jam, and one with just butter. I liked the cream one most. That cost him £9.00 by the time he'd had some himself.

'The local beach (Mawgan Porth) is nice and sandy and allows dogs all year round. It also has an estuary flowing in where sensible dogs like me who hate waves can have a paddle. That said, my daddy did get me to paddle in three inches of incoming seawater for the first time EVER yesterday. The waves come in as much as thirty or forty yards sometimes like mini tsunami, so he stood and waited and cuddled me when the wave approached and told me it was OK. I wasn't sure about it at all, but he was making such a pillock of himself in his soaked trainers that I didn't run for it.

'On the second full day, a good twelve people on the beach said something to the effect of "It's Max again!" On its own that's probably not funny to you humans, but it amused me in a somewhat bittersweet way because no one said "It's Bob again!" That's probably because no one asked his name to start with. Poor daddy.

'There were only a couple of downsides to the holiday. Firstly, the bathroom was rather compact. My daddy said he's had worse in Amsterdam, but the loo seat was strangely short, so basically you had to decide on the primary purpose of your visit before sitting down, as under no circumstances was a multi-function visit possible. Secondly, my daddy seemed sad at times and had tears in his eyes. I think that was something to do with the relative life expectancies of dogs and humans, but I don't understand that bit cos I'm a dog and I've only just learned to type.

'Still, chin up, at least I know my daddy loves me.

'Max xx'

32) I've had two wives. The first one is barking mad but begat me a family, while the second is normal and tends to my every whim. What has Pyror got to say about that? What happens in hologramatic heaven? Do I have to get along with both of them? I'm very confused.

Well if you're *really* lucky, it might just work out however you want it to!

There are, of course, a number of things that might affect the outcome (or indeed 'effect' it – ho ho – there's nothing like a good grammar joke), and largely they're a matter for your own conscious and theological conclusions, although feel free to believe the same as me if that helps, while at the same time remembering that it's your universe and no one has the right to *tell* you what to believe.

Maybe hologramatic heaven will be like one of the more common traditional models of paradise, where everyone is so all-consumed with singing Pyror's praises that they largely or even completely ignore everyone that they ever knew, or at the very least any worldly personal problems they had suffered or caused, so that they can praise Pyror all day every day for all eternity.

Personally, I don't think Pyror would want just that, not that a musical hologram wouldn't be *good* but if it could do even more that might be *better*. Also, quite frankly, the non-stop praising theory seems rather simplistic, some people might say it makes him sound like an egomaniac and, if I may be so bold, it sounds crucifyingly dull doing the same thing all day every day for ever, even if some people might think it better than working for a living.

So this is what I think.

I think that with a bit of luck, and I offer no firm guarantee because Pyror could always change his mind if he wanted to, that as a reward for participating in Pyror's Quantum String Theoriverse, once the competition has finished, each and every thing that ever lived (by which I mean things that exhibited the power of conscious thought, including guinea pigs and perhaps some of the, ahem, brighter plants) will be allowed to choose one or more of the $2^{\text{GOOGOLPLEX}}$ possible universes and to live in that forever complete, if they so choose and have at least tried to help Pyror score a point in the competition, with their very own top of the range flying saucer with force shields, invisibility cloaks and photon torpedoes in which they can whiz round the universe doing pretty much whatever they like, whenever they like, with whoever they like and to whoever they like including, if they so choose but *only* if they so choose, praising the preferred manifestation(s) of their god(s) all day every day.

But, if you do decide to go around zapping things in your flying saucer, there will be consequences.

If you kill a fish in the primordial soup and it turns out that you were related, it might be 'bad', in that if the paradoxes were allowed to resolve you would be erased from history. I therefore assume that paradise will come with an 'undo' button which you could use after being informed, via a special three-dimensional hologramatic screen in your flying saucer, through which you could also watch anything that ever happened in your own personal universe from any vantage point, and zoom in or out for any level of detail you wanted, of what the consequences of your actions would be if you went ahead with them.

Obviously if you killed the fish, hit 'save' and therefore erased yourself from history it would then be entirely your own responsibility, whether because you weren't paying attention or because you were fed up and wanted it to be game over, at which point your universe would probably be moved to the recycle bin, so technically you could still be restored unless you pre-invoked a macro that auto-hit the 'empty recycle bin' button once you were in it!

Likewise, in this scenario, if while laughing maniacally you zap your ex's great grandparents with a photon torpedo and hit 'save' it might well erase not only your ex but also any children you'd conceived, or at least erase them from your own personal universe; no doubt they would still be safe in their own little holograms. And while many parents fancy a break from their kids now and again, not too many want to completely erase them from history. So all I ask, while flying your heavily armed battle saucer, is that you think before vaporising your ex's entire family tree.

Similar challenges will doubtlessly also present themselves by virtue of the fact that you'll be able to choose how good looking, popular, intelligent, invincible and wealthy you want to be (if indeed you decide that money and/or possessions are required or even allowed in the heaven you choose). I would therefore certainly counsel against, for example, excessive promiscuity, particularly on your home planet, although obviously it'll be entirely your own choice (or rather it'll be your choice about the degree to which you get to choose things; you may if you wish decide to abdicate some or indeed all responsibility for making choices!).

Still, at least paradise should be interesting, it certainly sounds like fun, and we've demonstrated in *The Word of Dog* that this model of paradise is scientifically AND theologically possible. I'm quite looking forward to it once I've done all I need to on earth. I hope I can see my daddy and most of his extended blended family there, and I hope that they *all* want to give my tummy a really *good* rub!

33) I once had a crisis of faith and it wasn't much fun. If I listen to you, is it going to happen again?

No. Mainly because you will now be completely secure in the knowledge that whatsoever you believe, as long as it doesn't conflict with what you, Pingpongor, have already proven unto thyself, ie that anything you can possibly imagine happening inside the universe *could* happen, even if it probably won't, and that outside the universe (where the universe is also contained within the outside, thus allowing for Pyror and any special guests to be omnipresent if they choose) are zero, one, or more than one gods who are happy to communicate with you if you feel like it on whatever level suits you, whether it's a concentrated effort occasionally on a Sunday morning, five times a day on your hands and knees, or the long slow silent admiration of a sunset from a cliff top high above the ocean: as long of course as you aren't an asshole which, despite anything you may or may not have done before, I submit there is no longer any reason to be now you've almost finished reading *The Word of Dog*.

Just remember that other people can be right at the same time, even if they disagree with you: it just means they're happy in their own little universe and probably want to stay there.

The *really* important thing is that *The Word of Dog* <u>proves</u> that no one has the right to tell you you're wrong, because as long as you mean well you won't be breaking any house rules, so you won't be erased from history.

And it really *is* that simple!

34) I just read a bumper sticker which plagiarised and corrupted read, 'It's not about whether you manage to avoid the storm that counts, it's how well you dance before the storm arrives.' What have you got to say about THAT Mr Merrybear?

Wise words indeed, and an interesting question to boot! Let us discuss. In fact, if it's OK with you, I'll assume you're happy for me to answer in two parts. If you're NOT happy with that, try sending a quick email to last week and I'll see if I can change my answer before you read it ... I do apologise ... Apparently that was a bit 'cheeky' of me ... Although in practical terms, it does seem to be the ideal solution ...

Part One. Ah yes, there will always be a storm, at least in the parts of the omniverse most of us seem to inhabit, especially if the oncoming storm in this particular saying is a euphemistic analogy for death, in which case it would seem reasonable that the actual dancing referred to is a metaphor for life, in which case it would seem perfectly reasonable to at least strive to achieve the positive outlook symbolised in that euphemistic metaphor: viz, it's not whether you manage to avoid death that counts, because sooner or later the reaper will visit us all, but it's how you manage to live beforehand that's important (albeit from a purely earthly and therefore transient perspective).

Long yet strangely readable and poignant sentence alert.

By a remarkable, almost incredible coincidence, almost as if some sort of contrived tenuous link were being engineered, possibly by an outside-the-box influence that was seeking creative new ways to make inside-the-box a happier place to be, *Dance Before the Storm* is the title of the official biography of the background and early days of the popular folk-rock musical group The Levellers, most lauded for their protests against social injustice, and all forms of repression and oppression, in such timeless classics as *Cardboard Box City*, *Social Insecurity*, *Barrel of the Gun*, and *Battle of the Beanfield*, but also a rich cornucopia of such inspirational songs as *One Way* ('There's only one way of life and that's your own') and *Men-An-Tol* (inspired by a largely cat-free megalithic site of the same name near St Ives in Cornwall, which much to my chagrin is extremely difficult to access if you happen to be a Newfoundland), the name of the band derived from the eponymous movement formed during the English Civil War (that movement itself named after a label applied to rural rebels from the earlier popular uprising known as the Midland Revolt after common land was stolen from the people by the rich, in Mad-B's opinion not unlike the privatisation of British Rail which he would like to see put back to *exactly* how it was prior to the closure of the Somerset and Dorset Joint Railway on 7 March 1966 only with more diesel-hydraulic locomotives and in assorted 'old' colours), who believed in religious tolerance and that the legitimacy of the state is created by the will or consent of its people, ie that government should serve the will of the people rather than its own megalomaniacal self-interests, and who proposed radical agrarian reforms to benefit the majority and a 'world turned upside down' where the humble were exalted, thus arguably mirroring the now often sung Magnificat which, according to the Gospel of Luke, Mary proclaimed during her 'Visitation' with her cousin who didst beget John the Baptist and which set off a chain of events.

Back to normal.

Whether the band themselves have any particular religious sensibilities, I have no idea.

So, to summarise part one of this question, I recommend that you buy a copy of *Dance Before the Storm* and some Levellers CDs (or their electronic or vinyl based equivalents). *Levelling the Land* is a popular CD for newcomers to their music. They haven't paid me to say that, nor would I accept payment if they offered. Max Merrybear is NOT for sale. I'd let them rub my tummy though. The Levellers: Pyror's favourite band!

Part Two. In this part, I shall assume that you've just looked out of the window, that an actual storm is actually coming in the very near future, and that you just fancy a bit of a boogie before battening down the hatches, in which case my first thought was to suggest rather glibly that if that's what floats your boat, then go for it. However it seldom hurts to be mindful of others, so I'd probably suggest that if, for example, you have a black male Newfoundland life partner, that before strutting your funky groove thang you call him in from the garden, settle him down by the fire, and perhaps offer him a light snack of mushroom and goat's cheese risotto, ideally made with a ripe Capricorn goat's cheese from the Lubborn Creamery in Somerset, and perhaps a fine medley of porcini, shitake and portabella mushrooms, the latter ideally cut into small cubes or wedges for the texture, while the remaining ingredients can simply be sliced for taste and convenience. A little freshly ground black pepper wouldn't hurt either. Now that's what I *call* dog food!

35) If we wish to equate Pyror to a god of love, does it mean that he loves Adolf Hitler and people who rape and murder children, or does he vengefully send them to burn for all eternity in the fires of hell and, if so, does that really make him any better than the perpetrators he seeks to punish?

It's an age-old question, that's for sure, and a debate that may well continue to dog history no matter how precisely I answer, especially as each new generation, reticent of trusting its forebears and with new assholes to consider, asks itself the same question, and the answer inevitably evolves along with the language used to ask it: words are born and die all the time as I'm sure you know, so it would be beyond ridiculous to make a film aimed, for example, at a modern English-speaking audience in what you (almost certainly wrongly) thought ancient Aramaic sounded like 2,000 years ago, eg the controversial Mel Gibson film *The Passion of the Christ*. How utterly pretentious. That makes me a Mad Max!

Anyway, to answer your question, as discussed earlier, it's a fine line for any aspiring deity to tread between allowing the freedom of choice and freewill that make life interesting, and having a collection of subservient manikins who would doubtlessly bore rigid the judges of the Best Quantum String Theoriverse competition, eliciting not even a Commendation Rosette. You could make such a universe pretty and arty for sure, but not alive in any worthwhile sense, and the universe most of us seem to inhabit has a healthy dollop of beauty despite all its occasional harsh savagery, as long as you pause to take a close look now and again.

On the other hand, one might say that murdering innocent races (à la Hitler), children, puppies or whatever is 'a little bit naughty'. Pyror might well therefore strive to erase naughty people from history (or at least consign them to it, like for example Lilith) by tweaking the house rules and rerunning the omniverse, but I refer again to *the fine line*. Unfortunately, as long as a deity, whether Pyror or you, Pingpongor, allows free will, which as a god you *have to* if you're going to be adjudged any good, there are probably always going to be assholes unless, and I think some of you may by now know what's coming, there are $2^{\text{GOOGOLPLEX}}$ different universes, or some other really big number but not quite that big. If there are that many different universes, billions upon trillions of them might both exhibit freewill and not have any assholes.

BUT. And that *was* a big but ... Literally ... It was in capitals ... Boom, boom ... Sorry ... If there are no assholes, you could argue that there would be no *The Word of Dog*, as the latter discusses the former, so without the former there could be no latter, so the universe in question might actually be pretty dull anyway. Let's face it, as you already know, it's not that easy being God.

So, to summarise, there will probably always be assholes unless Pyror can find some future puppy capable of writing an asshole-free version of *The Word of Dog*, in which case he could then possibly arrange for all assholes to be erased from history and rewrite the past so that there never were any Adolf Hitlers, Genghis Khans, child murderers, paedophiles, and so that there would be no need for a hell or oblivion to punish or forget them in because they wouldn't have existed, therefore no eternally damnable crimes would have been committed. Ditto for many other forms of self-serving acts of wicked naughtiness against one's fellow man or dog. That could of course prove tricky, given that Genghis Khans lived long ago in the past, ie before the mass printing of books written by dogs that haven't been born yet was reasonably practicable, and at a time when dogs weren't generally apportioned the intellectual admiration they enjoy today. By all means email me any bright ideas you may have though.

However, *if* you're reading this and you're now a *little* bit worried because when you were ten years old you stole a spud gun and some football-related rub on transfers from Sansom's Hardware and Toy Shop in Wincanton, Somerset (cough, Mad-B, cough, Pyror told me what you did), and you're worried that you're going to be erased from history for being one of the many light-fingered mini-assholes who may unwittingly have contributed to the closure of the store, all I can really suggest is that you fling yourself prostrate before Pyror and ask him to erase just your *wrongdoing* from history, rather than your entire psyche and essence. And remember, traditionally to gain redemption you have to repent of your *misdeeds*, rather than just repenting that you've been *caught*. You wretched sinner.

Pyror will usually also take, 'Oops. Sorry. That was a bit naughty of me. I'll try really hard not to do it again,' as a suitable apology as long as you truly mean it. Pyror's nice like that. He's not a vengeful deity, despite what the sensationalist tabloids may have claimed in days of yore ... Although the whole flinging yourself prostrate thing is quite funny ... I'll let you decide.

36) Do you ever worry about the world becoming overrun by cats, Schrödinger's or otherwise, and what is the earliest point in the future when that might occur?

Oh goodness me, yes. I think that's something we've all worried about at some stage, the main concern being that, unlike dogs, cats can breed like rabbits!

Caramel, the oldest of the three cats with whom I live, does seem rather highly strung, perhaps because of the incessant procession of innocent bunnies and other critters she tortures to death and devours on the doorstep and the possible consequential nightmares that one day she'll meet a really *big* rabbit who wants to brutally avenge the crimes against his species, or perhaps a more regularly sized rabbit with nasty big pointy teeth not even remotely dissimilar to the murderous Rabbit of Caerbannog in *Monty Python and the Holy Grail*. However none of the cats in my blended family have much to say on the mathematics of feline reproduction, quite possibly because they've all been neutered. Therefore, as Mad-B's stepdaughter Abi is a Trainee Veterinary Nurse, I asked her whether she was able to shed any light on the subject.

Apparently, if you start with a single female kitten on the day it's born, it can get pregnant at aged four months, gestating for two months and therefore producing a litter of usually up to five kittens, despite the somewhat more generous litter sizes customarily experienced by aficionados and other associates of places named St Ives, all of which may or may not, but we have to assume *may*, be female. Of course, subject to the availability of one or more willing and able Tom Cats, not related to the female kittens both for the sake of moral decency and simplifying the calculations, those five kittens may then six months later give birth to five more female kittens each, along with the original kitten, herself now aged one year, having another litter of five. As you may appreciate, the burgeoning family tree rapidly becomes complicated, with each new and previous generation spewing forth ever more offspring in all directions. The rate of begetting really could be quite astonishing.

Mild nerd alert!

Luckily, I was able to calculate that the population appeared to obey the formula $(L+1)^G$ where L is the litter size, ie five, and G is the number of generations. Therefore, with a new generation every six months, after five years there would be ten generations, so $(5+1)^{10}$ cats, or 6^{10}, or 60,466,176 (so roughly 6×10^7).

Back to normal.

Rather worryingly, therefore, after 10 years there would be over three million billion descendants from just the one single original day-old kitten. And one or more very exhausted Tom Cats.

Unfortunately, and no doubt some of you with veterinary experience, or perhaps those of you living with a cat who's been a little bit too free with her love (and I don't mean the sort they often feign when they want feeding), may have spotted that while the initial (and subsequent) kitten would have her *first* litter at six months, she *could* get pregnant again just one month after giving birth, so producing extra litters at nine and fifteen months etc as well as twelve, eighteen and so on. Ouch! That *really* complicates things. And means a *lot* more cats a *lot* more quickly!

Naturally, I did try to model this cat population with a view to finding a generic formula based on time and litter size, but my publisher won't let me print my extended workings on the basis that it might offend anyone vaguely sane, especially as I haven't yet succeeded in merging the multi-generation-stream numeric patterns I was attempting to formularise with the usual formula for the sum to n terms ($S_n = (n^2+n)/2$) of a series of numbers, which means that I am currently unable to calculate the precise moment in 'time' (which we now know as something of an abstract concept anyway) when the entire land mass of the earth is overrun; the land mass being a little under 150 million square kilometres, so room for somewhere in the region of five thousand trillion cats.

One thing I have managed to calculate though, is that you probably wouldn't want to be the last mouse!

Anyway, budding mathematicians wanting five minutes of fame might like to email me. Usual conditions apply. For completeness, please factor in feline menopause at around twelve years and a life expectancy of say twenty years for simplicity's sake, albeit we can deduce, given that as three million billion (3×10^{15}) in the simple model is just one generation away from exceeding the five thousand trillion (5×10^{15}) needed for Armageddon, even in the simple model the earth would be overrun after 10 years 6 months, and in the more complex model it certainly wouldn't be any later!

It would, however, be remiss of me not to mention that I suspect the population of just the first two main legs of the family tree (in essence catering for the 'nine month' branch but not the fifteen month branch etc) without menopause and death, where 'S' is the number of three-monthly seasons, 'C' is the number of cats, and 'L' is still the litter size, would obey the Excel-inspired formula $C=(EXP(ROUNDDOWN(S/2,0)*LN(L+1))) + (EXP(ROUNDDOWN(((S-1)/2),0)*LN(L+1))-1)$ and that the answer would therefore be 10 years 3 months with a population now spewing forth into the sea of circa 7.31 thousand trillion cats, menopause and death therefore not affecting the answer, although with the extra (less populous) legs of the family tree factored in, I suspect that the answer will probably end up being 10 years exactly.

37) Would you pay £10,000 for a bottle of wine? In your reply, please quote all figures as at the time the question was asked, rather than attempting to compensate for future inflation to benefit any readers studying this work two thousand years from now.

That's an interesting but unexpected question at this stage, yet intriguingly one which allows me to comment on the frailty of human perceptions, and therefore perhaps encouraging at least some of my devoted readers to clear their minds of clutter and embrace some enhanced thinking and reasoning. In fact it's a splendid question. Well done!

As you already know, I am something of a gourmand compared to many other canines or species, although I would certainly defer to dolphins on the subject of piscine cuisine. Suffice to say, however, that in culinary terms I sometimes like nothing better than to chow down on a doggy bag from the two-Michelin-starred restaurant Le Manoir aux Quat' Saisons near Oxford (and I do see the irony that, if you'll pardon my French, the somewhat anglicised pronunciation of 'Quat' is 'Cat'!); not that Mad-B can afford to dine there, but I do have other gastronomic connections, such as my well-heeled friend Jenny who is somewhat prone to 'lunching'. Why exactly Raymond Blanc, fine chef though he certainly is, also feels unable to offer me fine canine dining in the actual restaurant itself I am unsure, although I that infer he may be in cahoots with the Dalswinton. After all, I am perfectly capable of resting my muzzle on the table and therefore eating from bone china just like any other diner, and I am hardly likely to rampage around the restaurant sniffing backsides, humping legs and begging for extras. That's more Jazz's territory. Still, at the time of writing, the most expensive bottle in their cellar was (allegedly – I have not myself yet been permitted to examine the wine list) a 1989 Petrus priced at just under £7,000.

Would I buy a bottle? No. Not at that price, and so probably also not by inference at the £10,000 about which you enquired, because I am very much conscious of the difference between *actual* value and *perceived* value (and indeed *transubstantiated* value where, for example, in some belief systems, something that looks like bread and tastes like bread is *actually* human flesh, whereas in my own personal universe it seems more likely that it was intended as a physical reminder of things outside the cosmic box of ping pong balls, and therefore perhaps more intrinsically valuable than a regular loaf but not necessarily *actual* flesh, although by all means believe what you like given that we've discussed energy transfer and therefore by implication matter transfer and so your opinion is perfectly valid whichever way you decide to look at it). And when a delicious 'Menu Découverte' at Le Manoir costs just £150.00 if the optional extras of coffee, petits fours and a cheese course are taken, personally I'd get far more pleasure from almost fifty complete meals than I would from a bottle of wine that in a blind tasting the large majority of people would get no more enjoyment from than with a similar £5.99 offering from Tesco's or, for that matter, a cold bottle of Stella Artois straight from the fridge. And if I asked Mad-B whether he'd rather have one bottle of wine or over 5,000 bottles of cold lager, I'm reasonably certain what the answer would be.

Furthermore, recently there was a case where a restaurant in New York was unable to sell a wine they'd priced at $90 a bottle. They put the price up to $100 and it started flying out of the cellar, not because it had changed, but because people unwittingly (or otherwise) now believed they were buying something 'extra special'.

On the other hand, it's just silly trying to tell someone they *don't* like having a filling at the dentist if they *do*, or that they *do* like sipping a piña colada on a tropical beach if they *don't*, so what is enjoyment if not the *perception* of pleasure. So while I personally wouldn't pay thousands of pounds for a bottle of wine, because I'd perceive that I was being taken for a ride, in your universe it may be a perfectly valid aspirational lifestyle choice, and it's *perfectly fine* that we see things differently because quite possibly for us things genuinely *are* different, so we have *absolutely* no need to bite each other as long as we begin our comparisons with, 'Really? In my universe ...' rather than 'Really? What an idiot. Where's my sharp stick ...'

38) I'm a little bit worried. My mummy's just died and my daddy died some time ago, so now I'm an orphan, although technically I am 48 years old, married with children and I've just finished paying off my mortgage. The problem is that my mummy's face keeps appearing ethereally to me far larger than life and with a look of primal horror from beyond the grave. Has my mummy gone to hell and if so is that where I'm going?

Oh my, no. At least not unless the shortcomings or behaviour of either or both of you has been unmitigable throughout life beyond redemption, ie so bad that there's nothing even Pyror can *reasonably* do about it without tightening the house rules so much that it messes things up completely for everyone else. There has to be an element of personal responsibility as well of course, so don't expect to get away with, 'Sure. I gassed six million people, but I'm practically a god so what's the worst that can happen?!'

However, regarding your mother's ethereal appearances, given that we've taken a look at possible energy transfer mechanisms, it doesn't mean that there isn't something transitional actually going on in 'the force' or 'ether', rather than you 'just imagining it'. The good news is it'll pass. The bad news is that her transitional 'spirit' may be horrified at something heinous you've thought, or done, or thought of doing, which she didn't think you were capable of, but because of her current energy transfer status, journey to the Pearly Gates if you prefer that analogy, she can now see in detail all the bad things that you, her precious little cherub, have ever thought and done.

Try to be a better person. Repent if you feel like it. Then she will be free. Even if it takes a while; even if it lasts a few generations, it's as nothing compared to life on Pyror's mantelpiece where, if you're lucky, your misdeeds will be forgotten because history will have been rewritten so that they (or most of them) never happened. Of course you don't have to believe that, but I submit that we've given a valid scientific and theological explanation that that's how it may be, so from this point onwards I further submit that it can only be counterproductive not to embrace *The Word of Dog*.

39) I'm fed up with being poor, hungry, humble, meek, merciful and lowly. How do I become as rich as Croesus, as full as Mr Creosote, as arrogant as Captain Air Cool (Hercule) Platini 'IQ Two-One-Two' from *Red Dwarf*, Series 5, Episode 1, *Holoship*, as extreme as Mother Teresa (oh come on – she was against contraceptives and ran a chain of soup kitchens and HIV hospices – am I, ie the person asking the question rather than Mr Merrybear or you, Mr Merrybear's beloved reader, the only one who thinks that was just a *little* bit suspicious?!), as merciless as Genghis Khan, given that apparently I didn't travel back in time and smite him, and as glorious as Caesar Augustus who, as we learnt in the previously mentioned utterly magnificent triumph of a drama *Rome*, was actually somewhat more successful in many ways than even his Uncle Julius?

How do you achieve all of that? According to Mad-B's son, Tom, it's a fairly simple matter of achieving world domination with a loyal army of robotic flying laser squid. Personally, I think, based on largely inferred information as it was before I was born, that Tom may have watched a few too many *Popeye* and *Teenage Mutant Hero Turtles* videos as a child (allegedly averring insistently at the time that those turtles were completely different to their American cousins of a similar name, viz *Teenage Mutant Ninja Turtles*, and that anyone unable to tell the difference was an unmitigated idiot). And perhaps a few too many episodes of the grossly underrated cartoon *Earthworm Jim*. Mad-B beats himself up about it frequently, although he insists that he did all he could to steer his son along the path of righteousness that was *Thomas the Tank Engine*, after whom his son was named.

The main question I have in reply is, 'Why would you want all of that?' If you achieve even one of those, except *maybe* if emulating Mother Teresa's less contentious traits, you're likely to be surrounded by nothing but fawning sycophants, many of whom will be wishing you dead and some of whom will certainly be musing on or fine tuning plans to actively hasten the speed of your demise. If you can take pleasure by living a simple life of moderation like, for example, a puppy, your chances of having it all taken away from you while worrying about the inevitability of your doom are greatly diminished. They will never be zero, Elizabeth knows that, but the more simply you feel able to live, the less distressing your existence is likely to be for everyone, the most important person in this particular case being yourself. It can be a really hard thing to get your head around, let alone live out, but personally I'd much rather be myself than Caesar Augustus or Genghis Khan, even on the odd days when Mad-B is away and I don't get my tummy rubbed.

So. To summarise. Either strive to be as happy as you can while 'playing the hand you've been dealt', and it isn't going to work for everyone, or start working on building a loyal army of robotic flying laser squid in the hope that it makes you a Happy Hitler. The choice is yours. It's your life. But don't for a moment think that an army of squid is necessarily the route to Utopia!

40) I work for a bank in a call centre. Almost everyone I speak to is miserable because we've charged them anything up to £30.00 to send an automated letter, costing us less than 50p to mass produce and post, regarding the fragile state of their account, thereby typically providing 25% of the bank's total profit, not that they like to admit that and they are happy to lie about it, but I've seen the confidential data and it sucks. I mean, they could all but wipe out child poverty in the UK with the bloated fat cat profits they make from screwing the poor. I'm depressed every day about the social misery I help to cause. I hate my life. I hate God for inflicting this pain on me. What have you got to say about THAT Mr Merrybear?

I confess I had to ask Mad-B for some advice on this one, as personally I have little experience of such places. He became somewhat agitated when discussing banks, for reasons previously touched upon, so a lot of what follows is based on my assumption that there may be some reason for his distress and the quotes he gave, and I confess at the time of writing it has made me slightly hesitant about the prospect of going to pay in a constant stream of hefty royalty cheques (so that I can put almost all of the money towards good causes, albeit perhaps somewhat idiosyncratically, such as creating jobs that pay living wages by opening ping pong ball and flying saucer factories). Anyway, based on the information I have available to me currently and notwithstanding any long-term benefits that may or may not ensue if the banking world evolves over time, and with Mad-B's observations in quotation marks …

Obviously 'no one actually *wants* to work in a call centre or for a bank', even though such institutions 'condition you through fear' to present so positive a public image that some people doubtlessly begin to believe the 'filthy stinking lies' of (specifically referring to a subset of banking executives rather than their downtrodden workers) those 'puss-sucking bastards' and therefore some people *perceive* that they actually enjoy going to work there.

You, however, by asking the question, have clearly not fallen foul of their 'egregious and unspeakable brainwashing': I commend you and recommend that you consider a change, although, despite my best wishes and even if you beseech Pyror, it may take a while. Here are some options:

- Check in your wardrobe and under your bed to see whether you have a fairy godmother. If you find one, ask her to wave her magic wand and maybe sprinkle you with a suitable magic powder or potion, such as Pixie or Fairy Dust, and perhaps a twinkling of Rainbow Stars.
- Do an Open University or similar degree over six years in your spare time and try for a new career. If you're on call centre wages it may even be almost completely free apart from having to buy some books, pens and paper.
- Become a ninja. I don't know why that might help, but in the absence of a loyal army of robotic flying laser squid Mad-B's son still seems to think it's the easiest way of achieving nirvana, which is quite ironic given the turtle thing. Perhaps aim to become a hero instead?
- Get another job. Before doing so, you may need to assess how much of your time you will spend on the phone

and whether the majority of the people with whom you will come into contact (if any) are likely to be happy or grumpy. You may or may not earn the same wages as well, but be aware that you *always* have a choice as long as you can live with the consequences, even if it means abandoning everything you have to risk and/or curtail your life to live on coconuts on the deserted island you sailed to on a raft you made from driftwood, flotsam and/or jetsam. Saying, 'I can't because ...' often gets misused to mean, 'I can't really be bothered,' or 'If I did that I wouldn't be able to watch EastEnders.' Be aware of these little voices that can rob you of the opportunity to try to make the changes to your life you'd genuinely like to, even though if like me you actually really just want to watch television sometimes, remember it is *your* life and it is *your* choice, and no one has a right to tell you otherwise, but please do try to make the best of the choices you make, even if that's one of the hardest things you may ever have to do.

Don't:

- Get another job for another financial institution doing the same thing and expect it to be any better: while not necessarily representing the views of Max Merrybear International, Mad-B avers that they all force their employees to believe they're 'different' and 'aiming to be the best', but in his humble opinion it isn't true and that they're all the same ... Although Mad-B does have a chum with pink hair who works for the Nationwide ... If you're desperate you could possibly try them ... I wonder if they're happy to take royalty cheques ...

Oh, and one minor observation, you cannot hate a being in whom you do not believe, so if you blame God/Pyror/Odin for your misery, you do nevertheless by implication believe in him, whether as none, one or many gods of whatever gender if any. I recommend that both you and he see what you can do to work towards a symbiotic relationship. You'll have to find a way to talk to him to do that of course, whether sitting on a quiet hill by the sea or in the clamour of a communal request on a Sunday morning. Whatever suits. Pyror likes his 'ping pong balls' to be happy when they talk to him, whether floating together or bouncing alone. And just in case you think it's inappropriate of me to tell Pyror what to do, in that I recommended you'd *both* have to work on it, it's fine. I'm allowed to suggest things to him. I'm his favourite puppy. He told me so when I was 'dead'. OK, technically he said 'equal' favourite puppy, but equal favourite is still favourite in my book; my book, as I'm sure you know by now, being *The Word of Dog* (and perhaps any future spin-offs and/or sequels). And he told me it's OK for you to make suggestions too.

41) Which came first, the chicken or the egg? Give your answer from a hybrid scientific and spiritual viewpoint and raise the bar when incorporating a paradigm shift in your blue skies thinking. Answer to my satisfaction and I shall warm to the idea of fervently believing everything else you've said.

Aha! A question which until now has confounded both human and canine philosophers for centuries, even though a team of well-meaning but misguided scientists from Warwick and Sheffield universities claim to have proved that it was the chicken, their profoundly inane reasoning being that there is a protein found in the eggshell that's only found in chicken's ovaries without, or so it appeared from the content of the broadsheet newspaper report, having considered that the protein in the ovaries might effectively have come from the shell, in essence by being replicated using the shell-based protein as a template. If that is the case, dog ... oops ... man, are they going to feel silly when they read this!

Anyway, there are actually (at least) two apparently different ways of answering the riddle depending on which sort of universe you inhabit, ie the degree of science or theology you prefer, although just to keep you happy I'll give a hybrid answer as you've requested.

The answer is this: 'A bucket of salt water that behaved according to the house rules Pyror laid down for whatever universe you happen to inhabit.'

You may think the answer a little odd, so naturally I shall explain in detail, but the main reason the answer probably seems a little odd at the moment is that the question is actually flawed.

Naturally, the Creationist viewpoint, and those of other similar 'Intelligent Design' based philosophies, is fairly easy to cover: 'The Lord didst create the chicken, which didst then layeth an egg, so the chicken just about came first, just like the team of scientists from Warwick and Sheffield universities has now proven, and The Lord didst then give man dominion over both the chicken and the egg and He saw that it was good. Amen!'

And, if you happen to live in a Creationist universe, that's fine.

The purely evolutionary mechanism is somewhat more tricky to reason: the simple fact is that you can't just start reproducing willy-nilly by laying eggs, nor indeed can you be an egg and just decide to change quasi-randomly into something like, for example, the first ever hen; it's a stupid idea and completely impossible in a 'normal' universe, even if you factor in Darwin's rather crass failure to allow for the ability of species to greatly influence the speed of their evolution through the *desire* to change.

Consider a bucket of salt water (and for anyone who gets turned on by largely mindless pedantry, we'll call it five litres of one molar sodium chloride solution). In fact, if you'll kindly indulge me for a couple of minutes, consider a bucket of salt water on your kitchen table next to your coins and boxes of ping pong balls and, for a small number of you, your double slit experiment and sack of Schrödinger's cats.

Leave it for a bit. Say a year or two.

What happens?

Assuming that you don't interfere with it, and your roof or pipes etc don't leak at all, and the house is adequately warm, it dries out; so your bucket of brine will have turned into a bucket containing solid table salt (just over 292.2 grams of it in the pedant's bucket). Fill it back up with water and it turns into brine again: it'll look the same for sure, although technically it'll be different water. Similarly, start with a chicken, get an egg, get a chicken from the egg and it may look similar when it's matured but it won't be the same chicken. Obviously it's not the *same* process: in many ways a chicken is a rather more complex arrangement than a (nevertheless miraculous) bucket of water, much in the same way that an omniverse is a rather more complex arrangement than a box of ping pong balls, but there's a conceptual similarity.

But what use is that? It explains nothing, at least not in any useful way. **Bear** with me.

Consider a bacterium. After a while, it'll split in two and reproduce (in itself an utterly astonishingly complicated biochemical process given that the 'wound' heals and the original bacterium generally survives). You then have two bacteria, normally identical for all intents and purposes, or at least so they would appear. But what if the bacterium is in a bucket of salt water that's been left to dry out on a kitchen table somewhere, and because the bacterium's environment is a bit 'weird' the act of splitting goes 'wrong', whether because of the intense saltiness or if there were some other ethereal force at work, such as a quasi-gestalt energy form or perhaps if you, Pingpongor, had breathed or sneezed 'the breath of god' into it.

What if we're left with a sort of Siamese bacteria situation where they stay co-joined? Or perhaps they're so co-joined that the 'child' bacterium is still subsumed within the cell wall of the 'parent' bacterium until the child manages to adapt to its environment enough to allow it to escape in a way that stings a bit but doesn't usually kill the parent. Therein might lie the evolutionary roots of reproduction in mammals (asexual at this point, so it would need more 'errors' before the mammals started forming romantic liaisons and expressing their love in a physical way). Or what if the 'child' is born before it's ready, although luckily the cell wall is thick enough to protect it while it finishes developing ie, in essence, due to the harsh environment the parent bacterium has involuntarily 'laid an egg', and then maybe the egg's DNA 'remembers' how it was born and does the same when it's ready to start a family. I submit that therein lies the evolutionary basis for a chicken and egg scenario, perhaps in a gradually warming primordial soup in a pre-Pangaean earth, in a really BIG bucket of salt water, ie the SEA, or maybe just some forgotten rock pool somewhere above the tide line three to four hundred million years ago, rather than necessarily here and now on a kitchen table. Although if bacteria *are* evolving on your kitchen table, it might be worth giving it a quick clean, especially if some of them have been possessed by Picasso's ghost. Use a bit of bleach if you like. After all, what's the worst that can happen …?

So, next time someone asks you whether it was the chicken or the egg which came first, I recommend that you answer thus: 'A bucket of salt water that behaved according to the house rules Pyror laid down for whatever universe you happen to inhabit, but you probably need to read *The Word of Dog* to understand why. Buy your own copy if you can. Mr Merrybear needs the royalties to help mend Boscombe roof.'

Chapter Twelve

Conclusions

42) So, if I understand it correctly, and I think I do because I, in this case the person reading the book, am an exceptionally gifted and talented individual and have therefore had the foresight to read *The Word of Dog* patiently, sequentially and conscientiously from beginning to end, feeling at least a comfortable level of understanding at each stage even if I ignored a bit of the maths, rather than just skipping to this section and expecting to find a simple answer to life, the universe and everything, the latter being an 'unfortunate' approach for anyone to take because they would have failed in their quest for enlightenment because of the synergistic and intellectual way you and I have built upon our knowledge at every stage, I now understand the universe/omniverse and what to do about it, which I wouldn't if I'd just skipped to this bit, so anyone glancing over my shoulder and reading this should avert their gaze immediately unless they want to get erased from history, not that I personally am threatening to erase them because that might make me an asshole, which label I certainly no longer deserve even if there may be one or two things from my past that with hindsight I might have done differently ; still,

what I *do* now realise is that the universe is not dissimilar to a box of ping pong balls, albeit rather larger and more flexible as well as more complicated, and rather like the air inside a ping pong ball which we don't generally need to worry about, matter is made up of a sort of quantum energy fizz (whatever quantum actually means) that nerds sometimes like to call 'energy strings', and because the universe is a bit more complicated than a box of ping pong balls and can even multitask like a woman, so little things like electrons can even behave like waves or particles at the same time, and it seems that the smallest bits of energy can probably even exist or not exist at the same time almost as if flipping a coin in a weird probability dimension, as weird as but different to what happens when I take a ping pong ball out of a box for a minute or two and draw a line around it zigzaggedly 'outside time' before putting it back, but I don't need to worry about that any more than I need to worry about the air inside a ping pong ball because ping pong balls are larger than sub-atomic particles and therefore I can make a set of rules about how my box of ping pong balls should behave and they'll generally behave accordingly (unless they want to get plucked out of the box, rived in twain, smote and cast into the dustbin of eternal nothingness by me, Pingpongor!), just like how Pyror, who in a sort of quantum jape can be one, none *and* many gods simultaneously and might have a few chums who act as demi-gods or saints as required, would seem to have an extremely complex set of rules in place for how the omniverse behaves, yet has deliberately allowed enough flexibility to allow life and freewill in what would otherwise just be an exquisite but lifeless work of art, the trade off for life and free will being that life doesn't come with any guarantees and there's usually some asshole who's trying to screw you over, but at least the universe *can* also offer all the excitement of a real firework display rather than just being like a somewhat lifeless photograph of a firework, although what we *do* about that is largely up to our conscience and our attitude towards the risk of embracing the excitement, plus any slight risk of being erased from history if we're unrepentantly or irredeemably *naughty*, plus any desire or otherwise to do what we think might help Pyror receive much kudos for winning a big trophy or rosette or something in the Best Quantum String Theoriverse competition we may well have been entered into ; and to summarise briefly, the gist of what I have inferred that you have been trying to convey to me is:

- You met Pyror when you 'died', but because humankind sometimes nails people to a tree for saying it'd be great if we all tried to get along, it seems that Pyror may have decided to have another go, only this time his messenger is a 12½ stone Newfoundland puppy, albeit that puppy avers strongly that he is not directly related to Pyror and is not here to be crucified in order to purge or re-purge the sins of man, original or otherwise.
- Science can tell us quite a lot about the universe; sometimes science is right; sometimes it changes its mind; often

it disagrees with itself; it can tell us quite a lot about how the universe behaves and probably behaved in the past, but *The Word of Dog* can also tell us why anyone actually bothered to have a universe in the first place. On its own not even a Large Hadron Collider can do that.

- Sometimes I've wondered if the entire universe is a figment of my imagination. Now I realise that's probably not completely true, even if I do sometimes live in my own little world. In my own little hologramatic heaven, however, I might have more choice about who's in and who's out.

- The universe is scary. We are not alone. 'Dark Flow' means that either there is something almost unimaginably massive outside of the universe that directly affects what happens inside, or that something even more scary is happening. *The Word of Dog*, howsoever theologically embraced, can calm our fears. And with a bit of luck the real Dr Who might be able to help as well. Captain James T Kirk would probably struggle. If I decide that it's all a bit *too* scary, I can hide behind the sofa and beseech Pyror to save me from the nasty space monsters.

- The lower my expectations of life the easier it is to be happy because it'll be more likely, although with no guarantee, that I'll get what I want, but it's probably OK if I still buy a lottery ticket and dream sometimes. Life is full of tradeoffs, ambitions and hopes, so if I don't have *any* expectations, even if my only aim in life is to live in near-perfect harmony with my fellow trees and lentils, I'm probably not making the most of things and I might have a bit more trouble choosing how my own personal heaven should look.

- Before establishing with any degree of certainty whether the chicken came before the egg, it always helps to work out in which universe you live, and whether someone is trying to be clever by asking you a quasi-paradoxical riddle, whether related to kits, cats, sacks, wives, or the evolution of omelettes.
- It probably won't hurt to clean the toilet once in a while and to eat healthily. Unless you mix bleach with acid, or spill your beans down the toilet and drink Domestos, in which case you probably weren't going to make old bones anyway ... Unless you're already old, in which case you can probably be philosophical about your ever more likely imminent 'onward journey' ... But generally people probably ought to take *some* responsibility for what they eat *and* the state of their toilets ... Even students.
- For some reason there is a thought now going through my mind, almost like in the worst kind of cheesy whodunnit where a new character appears in the very last chapter with a smoking gun and a guilty expression, that your favourite toy is called Mr Monkey, aka the 'Happy Pet Big Buddie Chucky the Chimp', a ten inch tall monkey with a 'realistic' voice box, and that every so often Mr Monkey gets ripped to shreds and has to go to 'hospital' for an 'operation', unlike Mad-B's teddy bear Edward with whom he used to play long division when he was four years old, then bear hunting with an airgun when he was fourteen, with Edward playing the part of the quarry, but who was never seen again after mysteriously disappearing at around about the time Mad-B's parents moved house when Mad-B was sixteen, the precise reasons for which disappearance Mad-B's

mother was unable to recall during her deathbed confessional and therefore took to the grave, although a skilled psychologist might infer that part of the enormous amount of fun Mad-B had while working at the hazardous waste transfer station was the presence of the adjoining landfill site, Edward Bear's most likely final resting place, and the consequent ethereal link to the halcyon days of Mad-B's childhood, now subsumed into his love for his 12½ stone Newfoundland puppy, Mr Max Merrybear, and the lady who enjoys cooking Max's porcini and goat's cheese saffron risotto. If we hadn't already answered everything we needed to know, I'd ask you what the heck those thoughts are doing in my head and how they got there.

- I will not had built a time machine but I might one day be reincarnated as a fictional television character from another universe, although probably not, unless perhaps I consciously decide to welcome a quasi-gestalt energy form into my Maxisphere. 'Ghosts' can only harass me if I let them.
- Now that we have used an informal conversational style, illustrated with very simple examples that I can model on a kitchen table with everyday items if I wish, juxtaposing humour with the most avant-garde aspects of quantum theory, to explain how science and various religions can be concurrently correct, conflict and war have become humanity's otiose shibboleths. Alas, due to the infallibilities of human nature, I suspect it may nevertheless be several years before the whole world learns to live in perfect harmony. Possibly even an entire decade. For some of us that can seem like a lifetime. For some of us, it is.
- If I were the sort of person to be offended when writers

use big words to massage their own egos without the meaning having an overt conspicuity based on the contextual positioning, ie what it means because of where it's at, I might have taken exception to the use in the previous bullet point of the words juxtaposing, otiose and shibboleths, even though this is *my* summary, so the choice of words was actually *mine*. Luckily I seem to remember that to juxtapose means to place close together, usually for contrasting effect, otiose means serving no practical purpose, and shibboleth means a custom, principle, or belief which distinguishes a particular class or group of people, especially a long-standing one regarded as outmoded or no longer important.

- Father Christmas exists. So does Dr Who. Until now, both have been somewhat misunderstood. There is a fictional television series based on the exploits of the real Dr Who. It is dumbed down a bit to make it easier to understand, but is nevertheless a noble and worthy feast of televisual entertainment.

- Given our discussions on the subject of the possible energy transference mechanisms of a person's psyche, it is not impossible that there is also a mechanism for a person or dog's dreams, on some occasions, to be a sequence of activities in which their ethereal spirit is actually participating in an alternative concurrent universe. Whether or not a person or dog should therefore apologise to their partner or significant others for their conduct in any such quasi-parallel universes, whether in their own dreams or the dreams of their partner, is probably a matter for each individual to decide for themselves on a case by case basis.

- Even Pyror can't help Elizabeth at the moment. We can but hope that in time he redeems her. Otherwise we're probably all boned. With a bit of luck part of her spirit is already in some sort of heaven, although that is a subject on which we would each have to form our own now-enlightened opinion, but it is possible. The same might apply, for example, to Alzheimer's sufferers, ie where part of their spirit quite clearly leaves the body a long time before final physical death. For now, however, all we can do is hope and have faith. Otherwise it's back to the boning thing. But for a caterpillar to become a butterfly, it has to go through the chrysalis phase, and often neither the caterpillar nor the butterfly feels entirely happy about the chrysalis. I might decide to reflect on that further at my leisure somewhen, because as a life-death-afterlife analogy I have a feeling that it's actually surprisingly profound.
- King James I allegedly once said of Richard Hooker that, 'I observe there is in Mr Hooker no affected language; but a grave, comprehensive, clear manifestation of reason, and that backed with the authority of the Scriptures, the fathers and schoolmen, and with all law both sacred and civil.' One might infer that history may say of Max Merrybear that, 'We observed there was in Mr Merrybear barely any affected language, but a clear and comprehensive manifestation of reason which was backed with the authority of Pyror himself, so as long as we're actively making an effort to help Pyror win the Best Quantum String Theoriverse competition no other law should be needed.'
- If a child tells you that he or she has actually seen an actual fairy at the actual bottom of the actual garden, it's possible

that there actually *is* an actual fairy at the actual bottom of the actual garden. If an adult tells you that they get attacked on a regular basis by fairies in heavily armoured wattle and daub battle saucers, it may be best to agree, smile, and back away slowly, *especially* as there's a slim chance that they might just be telling the truth.

- No conspiracy theorists should ever infer that the actual reason you, Mr Max Merrybear, want to replace Boscombe church roof is due to your guilt pangs over your possible involvement in the reburying of what was entirely possibly Richard Hooker's fibula, especially given the tendencies in bygone days for churches to share 'relics' such as bones of famous holy people as 'tourist attractions', so even if *most* of Richard Hooker's body was buried in Bishopsbourne in Kent, and by that time relic sharing was generally passé, it is still perfectly possible that his fibula was buried in Boscombe in the Bourne Valley where two Bishops live and where he wrote a pivotal multi-volume tome, just like some people want half of their ashes scattered on their favourite beach and the rest up their favourite mountain. Furthermore, no one should ever infer that you, Mr Max Merrybear, covertly dug the fibula back up and devoured it. So, to summarise, Mr Max Merrybear did NOT eat one of Richard Hooker's leg bones, nor indeed has he ever violated a Rector's leg.
- Good dogs go to heaven. So do good guinea pigs. Maybe even cats. If anyone thinks they've gone to heaven and no good dogs are present (not to be confused with no-good dogs, which is why it's useful to understand and occasionally employ the use of hyphens in compound adjectives

where appropriate), it may be best to check whether you boarded the right flying saucer. If you have to explain to Satan that you're in the wrong place, you might want to buy him a few pints of cheap cider and put *Scars of the Crucifix* on the netherworld jukebox to butter him up before raising the subject of obtaining a ticket for your return journey out of there. Or, depending on your chosen belief system, you could just (genuinely) repent of your sins and ask for Pyror's help in achieving redemption: not many people genuinely want to be violated non-stop by demons in a lake of fire and sulphur.

- Following on from our discussions regarding energy transfer mechanisms, what with brains using fizzing wavicles of energy to operate, such as electricity at a quantum level, technically there's no reason whatsoever why a brain might not discharge wavicles into the cosmic ether to be sensed by another receiving brain if suitably tuned to the transmitting brain, such as is occasionally alleged between twins, soul mates, or authors and readers when contemplating monkeys and teddy bears, and therefore also providing you, Max Merrybear, with a mechanism to transfer instructions into Mad-B's brain for subsequent typing with his opposable thumbs, so it's perfectly reasonable for a dog to attest to being the author of this work and Mad-B to be your typist/amanuensis or, if one were feeling magnanimous and wanted to aggrandise his job title while recognising that Pyror actually did all the important bits with a bit of help from a willing puppy, Head of Amanuensis Services.

- The universe might be bouncy. If it is that might provide a mechanism for 'remembering' things that happened in previous versions of the universe that in theory might or might not also still be ongoing in parallel concurrent universes, but are no longer widely accepted as fact because they sound weird to those of us who live here and now in the most popular universe, such as there being an Adam whose first wife was Lilith, but the Lord didn't see that it was good, so bounced/reran the universe.

- If the universe *is* bouncy and exists in an environment where things happen outside of time, and if we exercise due diligence and care, we can use the word 'forever' occasionally for the sake of brevity without worrying that mathematically we've given up or got the answer wrong, as long as we're careful how we use it and we've understood *The Word of Dog*.
- I seem to think that you want to offer me congratulations on my telepathy observations, noting that's exactly what you were thinking, and that I must have read your mind.
- By unifying theology with 'M-theory', which Professor Edward Witten came up with to itself mathematically unify various branches of string theory, we have now made theology accessible to scientists if they wish to access it. Perhaps somewhat ironically, although in its own way no less importantly, we have also made quantum mechanics more accessible to Bishops. Furthermore, by interfacing M-theory and Dark Flow with *The Word of Dog*, it seems entirely reasonable to conclude that there is a mechanism to transfer a person's spirit to outside of our universe leading up to or sometime after the final end of earthly life. Quantum mechanics therefore makes theology a scientifically valid concept. We should do our best to avoid being

nasty to each other over subtly different interpretations of what it all means however, therefore ensuring that the Head of Omniversal Services waxes gleeful at our harmonious acts of altruistic kindness.

- Just this side of heaven is a place called *Rainbow Island*. When a sentient being of whatever species finally dies their fully gestalt energy transfers there. As long as we have faith, that's probably a good thing. Once all the gestalt energy forms with which the sentient being in question had a symbiotic love-based relationship are gathered together on the island, they can transfer to a hologramatic heaven of their choosing. In the meantime each inhabitant of Rainbow Island can act, if they see fit, as a Guardian Angel to one or more Mortals of their choosing. Later, should one sentient being not wish to share a heaven with any gestalt partners they may have had, they will each covertly be granted a separate heaven where they can choose how the other is manifested unto them such that, for example, Mr Smith might appear to have Mrs Smith in *his* heaven, but in reality that particular Mrs Smith could be an exact replica, whereas the actual Mrs Smith might have had other ideas which are 'now' being fulfilled separately elsewhere. Likewise, their loyal hound could, if he/she so chose, have a third heaven where he/she owned a chocolate factory and a fine dining restaurant, while Mr and Mrs Smith had a replica puppy who looked and acted like the genuine hound in every way. It may even be that the puppy could have a separate universe but to drop in regularly (by transferring his/her energy into the replica puppy) to see how Mr or Mrs Smith's universes were going. That's the thing good about heavens; they need to be flexible, or one man's heaven could be another woman's hell.

- It is entirely reasonable for a Mortal to welcome some of their Guardian Angel's *actual* 'spirit' into their personal Maxisphere, so for example the Mortal's heart can retain an *actual physical presence* of their loved one, rather than just a memory, as long as both parties are willing participants in the energy transfer and receipt.

- It might help if you can avoid committing suicide in an attempt to hasten the reunion with your Guardian Angel: if it disturbs the ether and means the universe has to be rerun, and your Guardian Angel has to wait a hundred million trillion trillion trillion trillion trillion trillion trillion trillion trillion trillion trillion trillion trillion trillion trillion years for you instead of three score years and ten, it might make them sad, tetchy even.

- I should never waste my money on talking to a 'medium' (unless I want to) because if I want to talk to someone who's differently alive, I can do it myself. And I don't need to use a Ouija board. I can just have a chat like I would have had when they were on earth, albeit by using thought energy rather than sound energy, ie I should probably talk telepathically rather than with my mouth, especially if I'm using public transport. If appropriate, I'm free to carry on rubbing their tummy. Likewise, I can do that by transferring fizzing brainwaves rather than actually needing to rub anything with my actual hands in public places.

- It is entirely possible in this universe to create energy/matter from nothing. If it were *not* possible, there wouldn't be a universe worth mentioning here in the first place, albeit it may be that only an outside-the-box deity can facilitate this. There seems to be a rule that occasionally allows 'nothing' to split into a 'plus' and 'minus' parcel of energy that still adds up to nothing. It is entirely possible that the 'minus' parcel tunnels off into a quasi-identical anti-universe which therefore helps fuel the expansion of this universe by increasing the 'plus' energy here, which in canine quantum theology seems more likely than a 'plus' and 'minus' recombining into a single or double 'plus', although in theory if we factor in everything we've learned both situations could occur, possibly at the same time. This may explain why the universe is expanding faster and faster.

- Way back in Chapter Two we decided that we would rewrite Physics and that I, your beloved reader who is also the person responsible for Question 42, could take the credit. I've already done a pretty good job of that. However, in addition I've just concluded that based on Newton's Third Law of Motion, if for every action there is an equal and opposite reaction, if there's a Dark Flow there should also be a Light Flow. Light Flow allows energy from outside of the universe to travel into the universe. I have therefore just discovered new branches of quantum mechanics and astrophysics. Like Professor Edward Witten, I may therefore be even brighter than Einstein. I hope one day that everyone sees The Light.

- If I want to, when I get to hologramatic heaven I can choose to go around blasting my enemies with the laser cannons mounted on top of my heavily armoured battle saucer, or showering those I love with rose petals from my love cannon. It'll be *my* heaven, so my choice. Of course they'll probably get to have a heaven too, although any afterlife consciences in opposing versions of heaven will probably be switched off, so I won't necessarily actually kill or love my enemies or friends unless my loved ones want to share the same universe, but it'll seem perfectly real to me in my own little heaven, so there can be multiple concurrent heavens, not that in any case they'll be constrained by time as we know it, which means heaven(s) can be perfect for everyone, even if the current omniverse isn't. And it'll keep (the) God(s) happy. And I can probably do it 'forever' without getting bored, because I'll probably be able to switch my conscious off or recycle myself whenever I feel like it and start all over again. And I'll look great on Pyror's mantelpiece.

- I therefore conclude that we should try to be nice to each other without having to nail anyone to anything this time. Is that correct?

Yes, that is the word of dog. Thanks be to dog.

Printed in Great Britain
by Amazon.co.uk, Ltd.,
Marston Gate.